S0-BWX-181

The Control of Noise in Ventilation Systems

a designers' guide

The Control of Noise in Ventilation Systems
a designers' guide

edited by:
M. A. Iqbal, B.Sc., M.Phil., C.Eng., M.I.Mech.E., M.I.H.V.E.
T. K. Willson, Member of the Institute of Acoustics
R. J. Thomas, M.Sc., F.Inst.P.

Atkins Research and Development

London
E. & F.N. SPON

First published 1977
by E. & F. N. Spon Ltd
11 *New Fetter Lane, London EC4P 4EE*

Printed in Great Britain by
Woodcote Publications Limited

ISBN 0 419 11050 X

Distributed in the U.S.A. by Halsted Press,
a Division of John Wiley & Sons, Inc., New York

Library of Congress Cataloging in Publication Data

Iqbal, M.A.
The control of noise in ventilation systems.

Includes index.
1. Ventilation – Noise control. I. Willson, T.K., joint author. II. Thomas, R. J., joint author. III. Title
TH681.I68 1975 697.9'2 75-19238
ISBN 0-470-42812-0

Contents

Preface

In the building services industry the building services engineer usually has no need to concern himself with the mathematics behind acoustics as this is a task for a noise control specialist. But he should know the sort of acoustic calculations to perform at various stages of his project and where to go for more information when necessary. This, in essence, is the purpose of this publication.

Initially, it was prepared as a guide for use within the Atkins Group to provide the company's building services engineers with sufficient information to enable them to cope adequately with noise problems associated with all stages of their projects. The internal handbook—as it was then—achieved such a high degree of success, and so many copies were borrowed by friends outside the company that it was decided to publish it as a guide for general use of the building services industry.

Clearly, it is not the aim of this publication to be a textbook on acoustics, but merely to provide a general background and create the level of acoustic awareness that is essential from the conception of a major building services project through to final commissioning. In this respect, it is hoped that it will prove to be as much benefit to engineers elsewhere as it was to those within the Atkins Group. For those readers who may be unfamiliar with the basic principles of acoustics a brief summary of acoustic fundamentals and terminology has been included in Appendix 6.

The nomographs and other tables in the text from which calculations can be made have been presented in both Imperial and metric units. It should be noted that data based on outside sources has, in some cases, been extrapolated in order to make it compatible with the methods of calculation described.

EPSOM
APRIL 1976

Acknowledgement

Atkins Research and Development of Epsom, Surrey, would like to acknowledge the substantial effort of many of its engineers and those in other companies of the W.S. Atkins Group, during the preparation of this guide.

CHAPTER ONE

The design process

1.1 Introduction

It is commonly believed that the building services engineer designs his airconditioning system to meet all the ventilation requirements and can then simply pass over his design to the acoustics man who comes up with a silencer, and all the noise problems are solved.

Most of us who actually are involved in system design know only too well how wrong this belief is. How frequently, for instance, the acoustics man demands enormous amounts of room for his bulky silencers, or insists that the ventilation system designed is altered, or the building layout changed.

All too often the building services engineer can neither change his design nor allow the extra space, as either may involve structural or architectural changes. At that stage it may be too late for any changes anyway. Consequently, higher noise levels have to be accepted, or considerable time and money has to be spent on the changes.

Decisions, therefore, even at very early stages of the project, should be taken in the light of acoustic awareness. Unfortunately, it is generally believed that to do any acoustic check or analysis, detailed information about the equipment and layout is necessary, as all acoustic work is supposed to involve elaborate calculations. This, certainly, is not always so. Detailed calculations are necessary, but only in the final stages of the project. At some stages simple checks giving approximate values are sufficient, while others require no calculations whatsoever. All that is necessary is fundamental acoustic awareness and simple 'rules of thumb'.

The flow diagram in Fig 1.1 illustrates the stages of a typical project and refers to the sections which describe the depth of treatment required at each stage.

1.2 Project conception.

At this stage

The amount of acoustic treatment necessary for a building is governed by the way that building is designed and laid out.

FIG 1.1 The Design Process

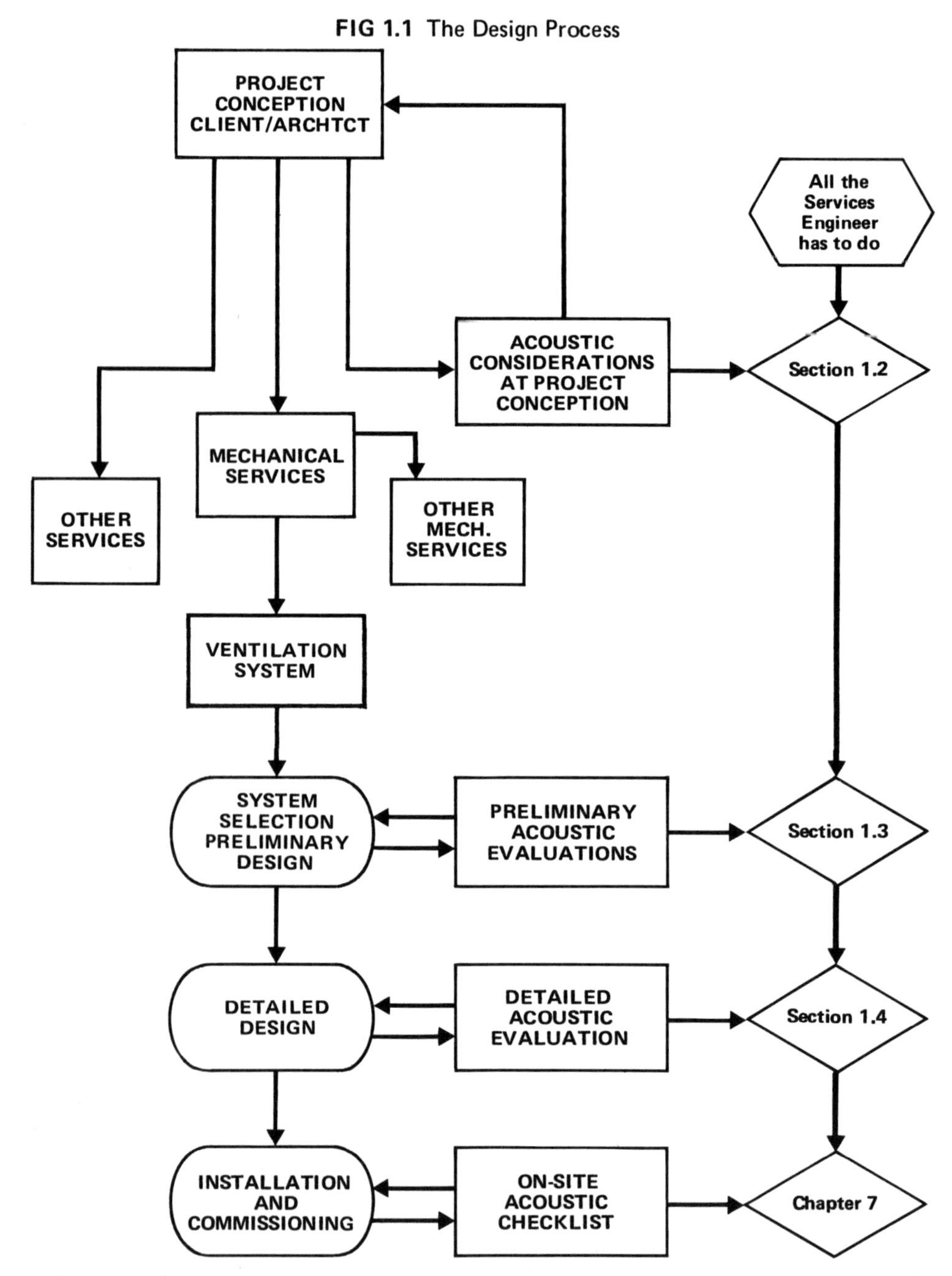

What to do

Apply this six-point check plan:

(1) Ensure the building is laid out with regard to noise insulation. An attempt must be made to separate the noise sensitive areas from the noisy areas. The noise sensitive areas should preferably be surrounded by reasonably quiet ones. As an extreme example, workshops or plant rooms should not be placed immediately adjacent to, or above or below, areas like conference rooms.

(2) Ensure that there is adequate noise insulation for rooms with low NC requirements. An office with a single glazed openable window looking out on to a busy street may result in an environment of, say, NC 40. Any amount of silencing of the ventilation system will not bring it down to NC 30. Adequate noise insulation must be provided to reduce the effect of other noise sources.

(3) Ensure there is enough sound absorption within the rooms. For most purposes in offices, banks and other commercial buildings the reverberation time should not exceed 1 to 1.4 s. To provide optimum reverberation times, the amount of absorptive treatment (absorptive ceilings, soft furnishing, etc) required has been found to be of the order of 6 times that present in untreated rooms, and can result in up to 8 dB reduction of noise level regardless of whether the noise source is inside or outside the room. In most cases, the addition of absorptive material will result in an improvement in the acoustical properties of the room.

(4) Ensure that there is adequate space allowed for the plant and ductwork. Almost all the problems encountered in acoustic installations are due to lack of adequate space. This clearly indicates that one of the first items to be considered by the building services engineer is the availability of adequate space. The lack of adequate space could lead to the placing of too much equipment into too small a place or to the expectation of higher performance from small equipment, hence utilising the equipment beyond its design capabilities. Particular attention, therefore, must be paid to the plant room size and the space available for ducts and silencers. As an approximate guide 15 per cent (ideally it should be 20 per cent but this may not always be possible; it should never be less than 10 per cent) of the total floor area served by the plant should be allowed for the plant room and one dimension of the plant room should be at least 11 m (35 ft) regardless of the plant size. In addition to this, space equivalent to 1 per cent of the floor area served by the duct must be allocated for the cross section of the supply and return ducts (ie ½ per cent for supply and ½ per cent for extract).

(5) Ensure correct location and construction of plant room. The position and the construction of the plant room are critical. Plant rooms should be away from noise sensitive areas and must be constructed of at least 225 mm (9 in) thick brickwork, either pointed on both faces or pointed on one face and plastered or rendered on the other. When doors are necessary in the plant room structure, these should be of at least 35 dB attenuation and supplied by an approved acoustic door manufacturer. These doors must be capable of being sealed adequately when shut. The plant room floor should be of at least 200 mm (8 in) dense concrete. When the plant room has to be over critical occupied spaces, an isolated imperforate suspended ceiling of at least 25 kg/m^2 (5 lb/ft^2) density will be required. The cavity between the soffit of the plant room floor slab and the suspended ceiling should be at least 300 mm (12 in) and must be provided with a 50 mm (2 in) thick layer of mineral wool or glass fibre.

(6) Obtain noise levels of the neighbourhood. In order that the building services can be designed without adding to the neighbourhood noise levels, it is essential that exact levels existing in the vicinity are obtained by actual measurements.

Table 1.1
Summary

Plant Room	
Minimum overall space	15% of floor area
Minimum single dimension	11m (35 ft)
Walls: minimum thickness	225 mm (9 in) brick
Doors: minimum attenuation	35 dB
Floors: minimum thickness	200 mm (8 in)
Ducts	
Suspended ceiling: minimum density (if between a duct and a critical space)	25 kg/m^2 (5 lb/ft^2)
Duct risers: minimum space	1% of floor area
Duct transitions: maximum core angle	30°

1.3 Preliminary acoustic evaluation.

At this stage

(1) The ventilation system is usually being sized and laid out.
(2) The fan volumes and pressures are known approximately, but the exact fan is usually not yet selected. Ductwork details may or may not have been established.
(3) It is generally desired to know:

(a) The approximate sizes of the silencers on the room side, so that adequate space may be allowed for them in the layout.
(b) If any silencing is required on the atmospheric side of the fan (by considering the noise levels of the neighbourhood), and if so, the approximate size.

What to do

Make sure that the duct velocities (see below) lie within the acceptable limits of the design, and use the method described here to arrive at the approximate silencer sizes.
(1) The first step is to get an idea of the fan sound power level (the terms PWL, power level, sound power level and SWL all have the same meaning and unless otherwise specified a reference of 10^{-12} W is implied: see Appendix 6 page 91). The manufacturer most likely to supply the fan may be able to provide this figure, but an indication is available from the nomograph Fig. 1.2. At this stage a single figure at 125 Hz is adequate. In the majority of the cases this band is the most critical, and selection on this band alone will ensure satisfactory levels in all other bands.

EXAMPLE:

Estimate roughly the power levels in the 125 Hz band of an axial fan delivering 5000m^3/h (3000 ft^3 /min) against 500 Pascal (2 in wg) pressure, operating at a reasonably efficient point on the curve.

A figure of 70 per cent for fan efficiency may be chosen as a suitable operating point. The example is shown on the nomograph in Fig. 1.2.

The fan power level therefore is 90 dB at 125 Hz.

(2) Enter the fan PWL (at 125 Hz) on the nomograph in Fig. 1.3(a).
(3) Draw a line joining the fan PWL and system attenuation and mark off where it cuts the reference line.
It may not be possible to estimate the exact system attenuation at this stage, so assume this to be 15 dB. In the majority of cases, this has been found to be within the range 12 to 17 dB in the 125 Hz band.
(4) From the reference line draw a straight line joining the design NC level, and extend that to arrive at the basic silencer insertion loss.
(5) Apply the room volume and air volume corrections given on the nomograph, to obtain the corrected silencer insertion loss; then, use the table to obtain the silencer length (Fig. 1.3(b)).
(6) Allow a 300 mm (12 in) length of silencer even if the calculated silencer insertion loss comes to 0 dB or just below. Assume that no treatment is necessary only if this figure is less than –5 dB.
(7) Allow for the silencer cross section being 2½ times that of the duct, and ensure that the total cone angle of the expansion piece does not exceed 30°.
(8) For noise to the exterior, refer to the nomograph in Fig. 6.1. This will give an indication of whether the effect of noise emerging from the fan inlets and outlets to the atmosphere is acceptable in a given neighbourhood noise level. Exact neighbourhood noise NC levels should be established by actual measurements.
Working this nomograph backwards will given an approximate insertion loss (in the 125 Hz band) of the silencer required on the atmospheric side.

FIG 1.2 Nomograph for Fan PWL estimation

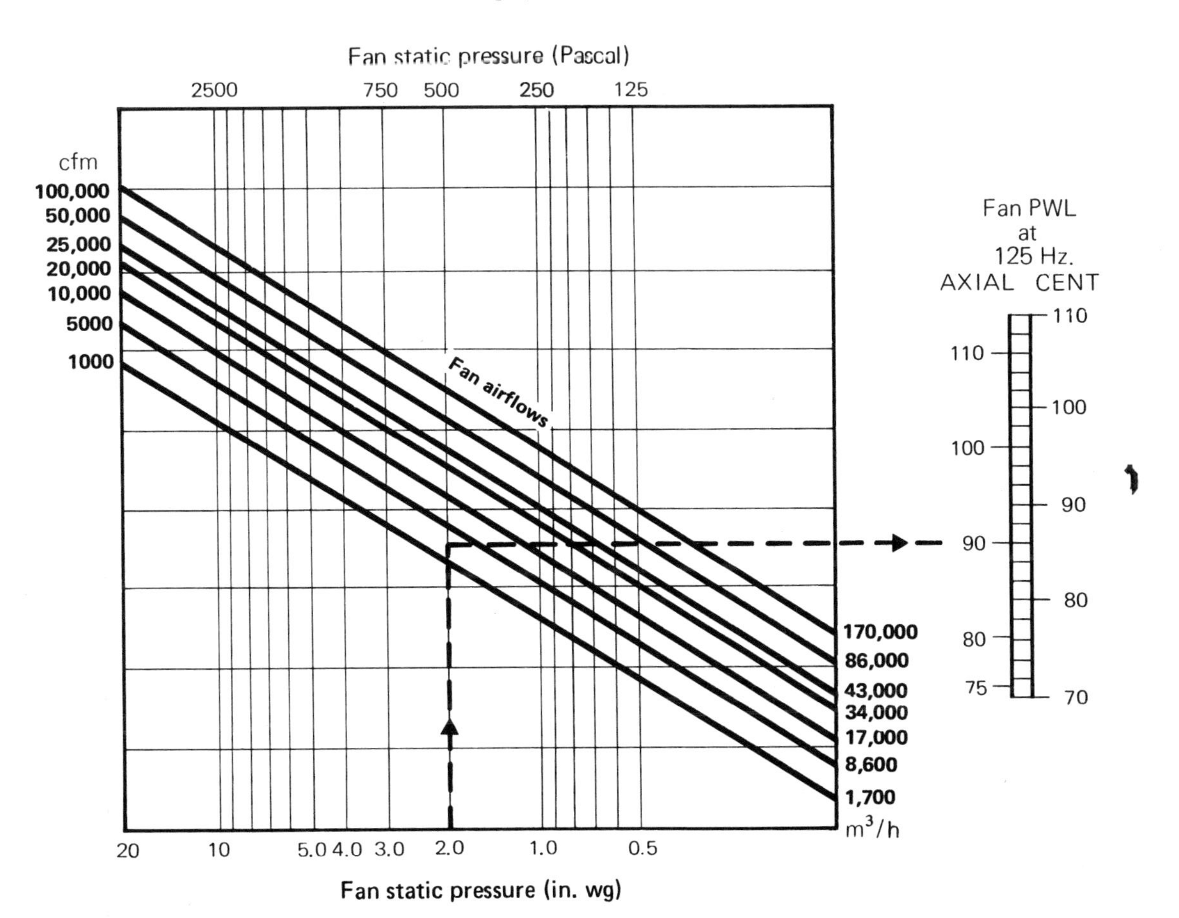

1) Corrections for fan efficiency (if known):
For every 10% above 70% subtract 2dB
For every 10% below 70% add 2dB

2) Spectrum correction add:

OCTAVE BAND Hz	63	125	250	500	1000	2000	4000	8000
AXIAL	+1	0	+1	0	–1	–4	–9	–15
CENTRIFUGAL	+2	0	–3	–4	–11	–16	–21	–26

FIG 1.3 (a) Nomograph for Preliminary Silencer Selection

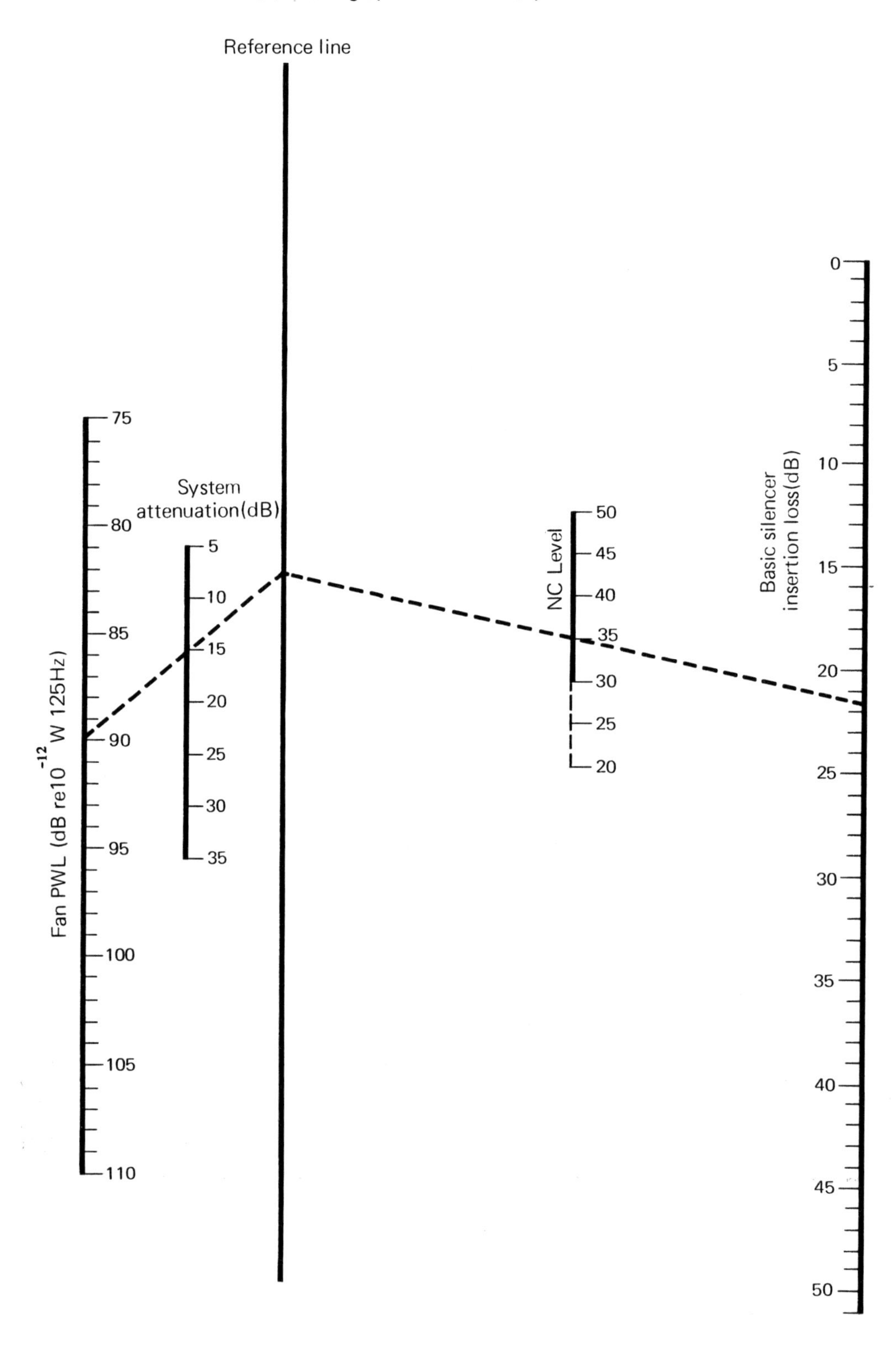

FIG 1.3(b) Corrections to Basic Silencer Insertion Loss

Correct silencer insertion loss = Basic silencer insertion loss + RVF + AVF

ROOM VOLUME FACTOR

m^3	ft^3	RVF
14	500	+3
28	1000	0
56	2000	–3
112	4000	–6
225	8000	–8
450	16000	–12
900	32000	–15
1800	64000	–18

AIR VOLUME FACTOR

P*	AVF
100	0
50	–3
20	–7
10	–10
5	–13
3	–16
1	–20

*P is percentage of the total fan air volume that reaches the room

SILENCER LENGTH

Correct silencer insertion loss	Length of silencer required	
dB	mm	in
–5 to 4	300	12
6	600	24
7	900	36
9	1200	48
10	1500	60
13	1800	72

(9) Establish the length of this silencer by using the table in Fig. 1.3. Comments regarding the cross section areas, and cone angles referred to in (7) also apply here, and these must be taken into account in the preliminary layout.

Velocities

The method described is applicable only to low velocity systems, where the maximum limits of velocity should be:
10 m/s (2000 ft/min) in the main risers.
7.5 m/s (1500 ft/min) in the branch ducts.
4 m/s (800 ft/min) in the ducts serving the grilles.
At these velocities the nomograph in Fig. 1.3 can be used safely for NC 30 and above. If, however, levels lower than NC 30 are required, then these velocities should be further reduced by 33 per cent for NC 25 and by 50 per cent for NC 20.

Where velocities in excess of 10 m/s (2000 ft/min) exist, the nomograph should not be used. At these velocities the HVCA recommend high velocity ductwork. This is significantly different from the conventional low velocity type, and is based on good airflow design principles, which allow for smooth airflow paths. A situation commonly encountered is the use of velocities in excess of 10 m/s (2000 ft/min) in conventional low velocity ductwork. This presents considerable problems due to velocity generated noise. In such situations attempts must be made to reduce the velocities to as low levels as the economics and the design can permit. Even then, if these are in excess of 10 m/s (2000 ft/min), the duct design must be modified to a high velocity type (see Fig. 1.4). Advice should be sought from a consultant regarding the special aspects of high velocity silencing techniques.

When laying the system out

(1) Avoid obstructions in the duct, particularly if the velocities are on the high side. If any obstructions are absolutely necessary, ensure that they have an aerofoil section.
(2) Duct components that are likely to cause turbulence (e.g. dampers) should not be placed close to bends or any other component that would create turbulence. A minimum separation of six diameters should be allowed.
(3) The use of dampers for balancing the system should be kept to a minimum. However, where this is necessary the dampers should be situated as far away as possible from the rooms or from the diffusers. It is better to install two sets of dampers each 20 per cent closed than one set 40 per cent closed. The noise generated by the two sets will be less than the noise from one highly restricted damper.

1.4 Detailed design stage.

At this stage

(1) The ventilation system has usually been designed and drawn up.
(2) The fans have been sized, and exact PWLs are available.

What to do

For a complete analysis, the following six steps are required:

(1) *Ductborne noise analysis.* Detailed calculations to analyse the effect of fan noise carried by the duct to the rooms, and evaluation of the minimum insertion loss of the silencers are necessary. A full octave band analysis is required to ensure that all the frequencies are checked. This is necessary only for one or two chosen branches that are likely to be critical. The fundamental method is described in Chapter 2.

FIG 1.4 Modifications to ductwork for high velocity systems

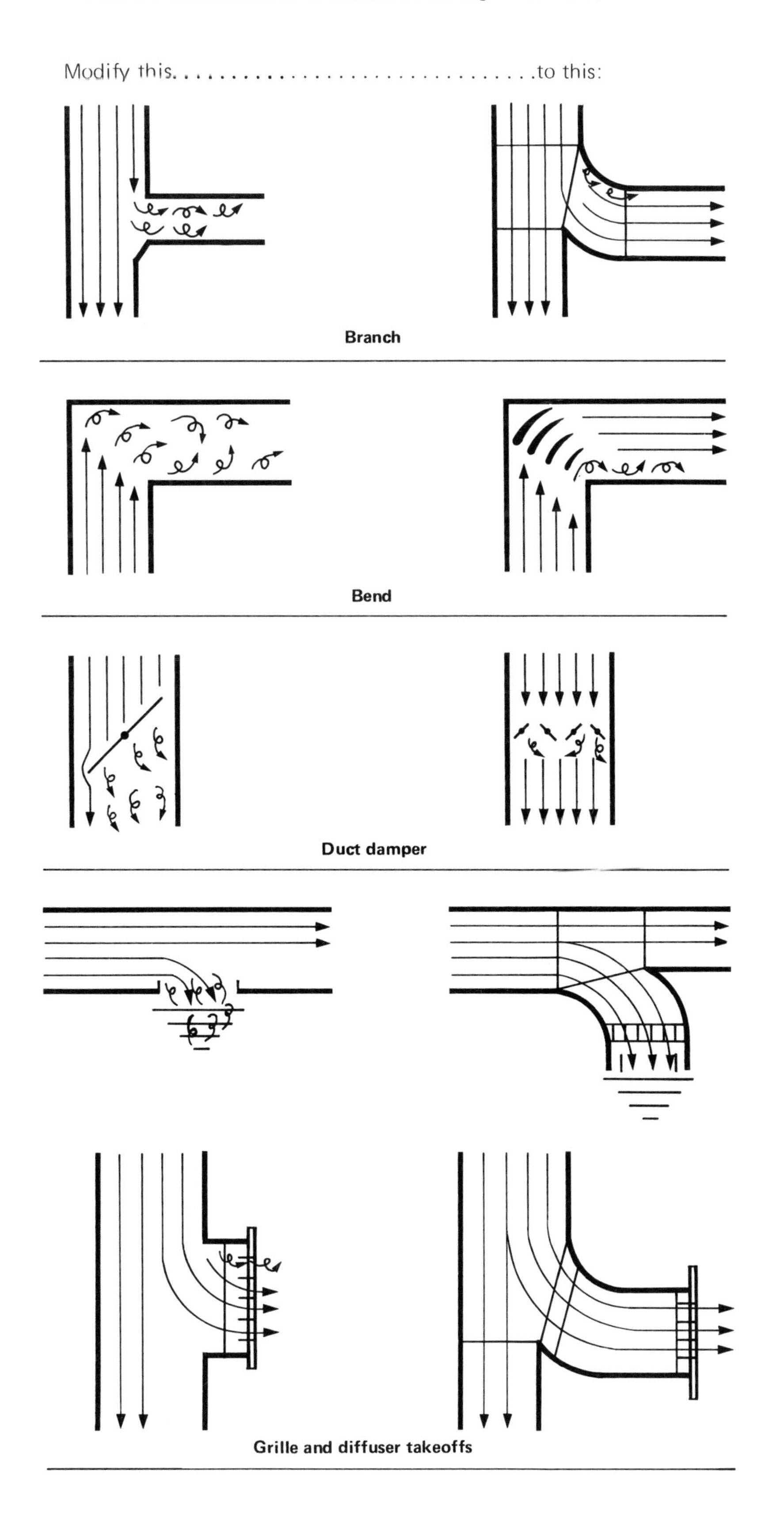

(2) *Velocity generated noise analysis.* A check is necessary to ensure that the noise generated due to air velocity in the various duct fittings is within acceptable limits. This again should be done over the full octave spectrum. The fundamental method and data are described in Chapter 3.

(3) *Duct breakout noise analysis.* Full octave calculations are necessary, to ensure that the noise breaking out from a duct passing over a noise sensitive area does not contribute to the noise level. If it does, then the evaluation of a suitable silencer is necessary to reduce the induct level. The basic method is described in Chapter 4.

(4) *Analysis of noise breakout to exterior.* This is necessary to ensure acceptable noise levels in the neighbourhood (see Chapter 6).

(5) *Cross talk elimination.* An evaluation of suitable silencers is required in order to eliminate the transmission of speech and office noise via interconnecting ductwork. Detailed calculations are generally not necessary. A simple selection method is described in Chapter 5.

The above steps should be carried out for both extract and supply systems, treating each as a separate system.

(6) *Vibration elimination.* In addition to the five noise reducing steps, elimination of vibration is also required. All rotating and oscillating machinery must be mounted on suitable isolators and provided with flexible connections where necessary.

Computer programs

Items (1), (2) and (3) are simple but involve lengthy calculations and frequent reference to tabulated data. This makes them ideally suited for digital computers.

Simple programs have been developed which are described in Appendix 5.

CHAPTER TWO

Ductborne noise

2.1 Introduction.

The basic problem is that of establishing whether such noise as may be generated by the air handling equipment will create objectionable noise in the occupied parts of the building and, if so, of formulating some means of reducing it to an acceptable level.

For this it is generally necessary to:

(1) Identify the noise source and establish its characteristics.
(2) Establish the critical duct branch or branches.
(3) Estimate the sound pressure level resulting in the most critical room or space.
(4) Establish the noise criterion in that room.
(5) Compare (4) and (3) and select suitable silencers if (3) is higher than (4).

The method of calculation is basically identical to that described in various publications, e.g. the IHVE guide, *Design for Sound* or the ASHRAE guide, except that greater emphasis has been placed on velocity generated noise, and duct breakout noise, about which little has been known until recently.

2.2 Identification of the noise source and establishing its characteristics.

The fan is the prime source of noise in any ventilation system. A small part of the horsepower supplied to the fan is radiated out as sound power. The higher the power supplied to the fan, the greater will be its acoustic power. The fan, therefore, must always be selected to operate at the point of maximum efficiency. This will also prove to be the optimum point acoustically.

For stable fan operation, it is important not to select a fan which is too large for the application. The design engineer should choose the working point on that part of the fan performance curve where there is assurance of stable fan operation despite small changes in system resistance. Proper sizing of the fan is one of the most important factors in ensuring a minimum of fan noise.

Almost all leading manufacturers give the sound power output of their fans, along with pressure, volume and horsepower characteristics. It is, therefore, possible to estimate the sound power output at any operating point. This is usually expressed as a single figure, and correction figures are generally supplied for each frequency band, so that power level in any octave band may be established.

Some manufacturers, however, still express the noise of their fans in sound pressure level. This is meaningless in this context and the engineer must be careful not to use it, unless sound pressure level is converted to sound power level.

2.3 Establishing critical duct branch or branches.

Having established the characteristics of the fan, the next step is to examine the duct systems served by each fan in order to evaluate the silencing required. The general method is to select a silencer capable of meeting the NC requirement of the most critical zone and to place it in the main duct just downstream of the fan. This will obviously oversilence the less critical zones but is usually the most satisfactory arrangement. Separate cross talk silencers should then be calculated and installed in the ducts interconnecting noisy and quiet areas to eliminate any cross transmission of noise. There are a few other approaches but the one described above is the quickest and simplest to use and should generally be adopted. As the reader gains design experience he will discover the other approaches himself and will be able to try them out.

The method of silencer selection described here caters for the most critical zone, which is not necessarily the quietest. A zone at the end of a long duct run with several bends may have sufficient attenuation within the system and require no further silencing. On the other hand, a less sensitive zone close to the fan or carrying a greater percentage of air or discharging into a smaller room may need additional silencing. For calculation purposes, therefore, a simple method is to consider only the branch leading to the first outlet unless there are subsequent downstream outlets which are:

(a) handling a greater air volume; or

(b) serving a more critical zone (i.e. a lower NC acceptability level).

If this is so, then calculations should be repeated for those branches to establish the maximum silencing required.

EXAMPLE 1:

Calculate which branches of the system shown in Fig. 2.1 are the more critical.

SOLUTION

(a) Branch AJKL discharges into the most critical area (NC 35) and therefore should be analysed.

(b) Branch AB, although not discharging into the most critical zone, has a relatively short length of ducting. It is also carrying the highest of the three air volumes (2074 m^3/h, 1200 ft^3/min). This must, therefore, be examined.

(c) Branch AJKT is further downstream than AJKL, is discharging into a less critical zone (NC 45) and is carrying the smallest air volume. This branch, therefore, can be safely ignored.

2.4 Estimation of sound pressure level.

The estimation of the sound pressure level resulting in a room involves four calculations:

(i) sound power level entering the room;

(ii) reverberant sound pressure level;

(iii) direct sound pressure level;

(iv) sound pressure level resulting from (ii) and (iii) taken together.

FIG 2.1 (metric) Ventilation system

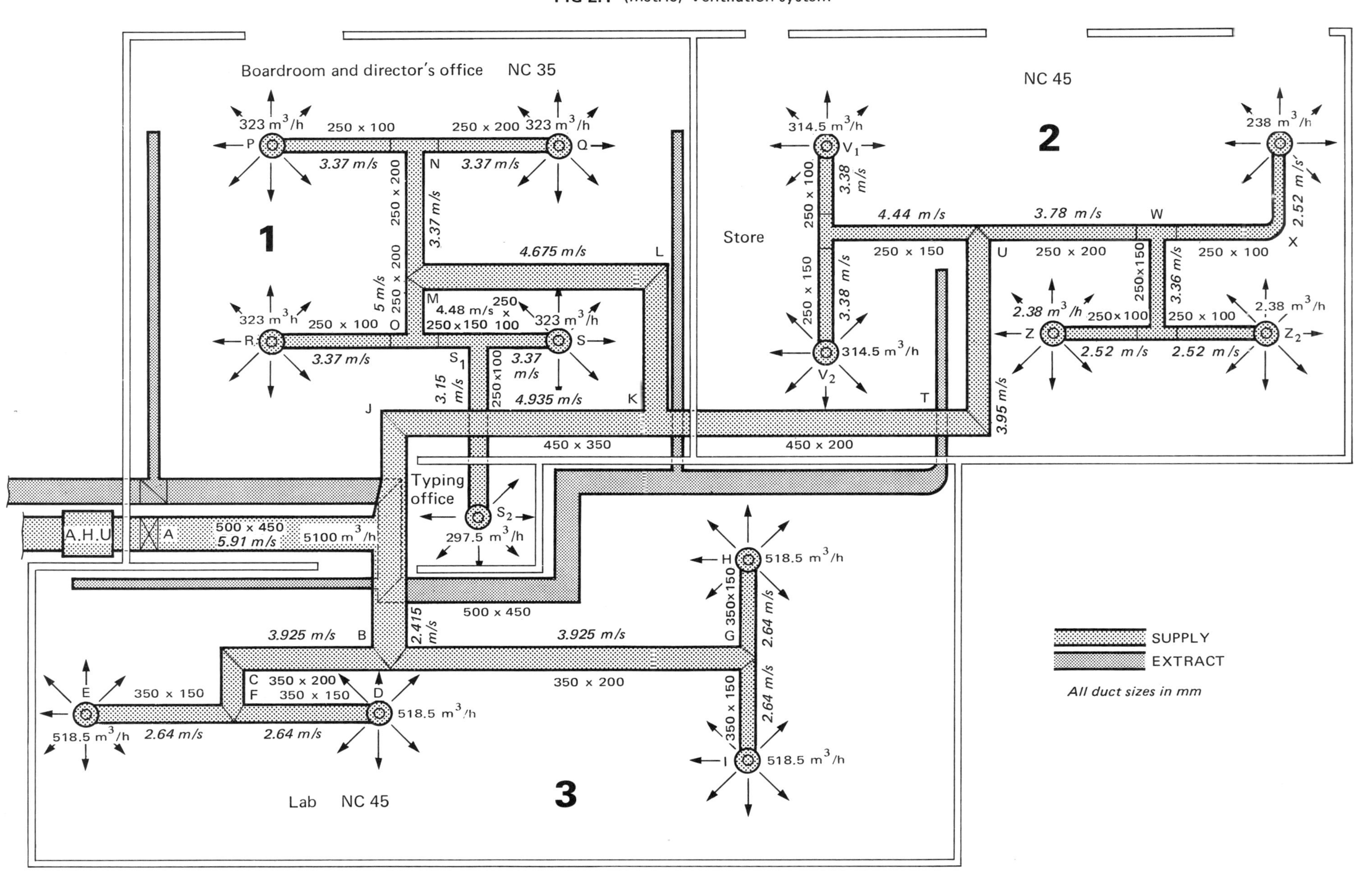

2.4.1 Sound power level entering the room

This is arrived at by subtracting the attenuation of all the ducts, bends and outlets from the sound power level of the fan.

The power level entering the room therefore is:

$$PWL_e = FAN.PWL - a - b - c.$$

Where PWL_e = power level entering the room

a = the attenuation of all the ducting in dB (Table 2.1)
b = the attenuation of all the bends and elbows in dB (Table 2.2)
c = the attenuation of the outlet in dB (Table 2.3).

Note

(a) Unlined ducts provide a small amount of attenuation which can be increased by the application of a suitable lining material. Table 2.1 gives the attenuation per foot (0.3 m) of various sizes and types of ducts.
(b) Bends generally provide relatively poor sound reduction at the lower frequencies.
(c) Lining with a suitable absorptive material, placed on the room side of the elbow where repeated reflection takes place, improves the attenuation. Table 2.2 gives the values of attenuation for various types of elbows.
(d) Outlets give almost negligible attenuation at high frequencies, i.e. all the power in the high frequency band is allowed to pass through to the room. For the low frequencies, however, some of this is reflected back and absorbed, thus, substantial low frequency attenuation is achieved. Attenuation is dependent on the total area of the outlet. Table 2.3 gives the attenuation of opening areas.

EXAMPLE 2:

Calculate the system attenuation for rectangular sheet steel ductwork consisting of: 3.6 m (12 ft) of 450 mm (18 in) duct; 6.6 m (22 ft) of 350 mm (14 in) duct; 9 m (30 ft) of 200 mm (8 in) duct; 2.4 m (8 ft) of 100 mm (4 in) duct; two 90° elbows at 350 mm (14 in); three 90° elbows at 200 mm (8 in); and a 0.06 m^2 (100 in^2) outlet area. From these estimate the power level entering the critical room, using the fan power level given in Example 7, page 25. (Only the smaller dimension of each duct cross-section is shown as this determines the attenuation).

SOLUTION

	63	125	250	500	1000	2000	4000	8000 Hz
Fan sound power level spectrum (dB re 0.0002 dyne/cm^2)	82	84	84	82	81	80	76	70
From Table 2.1, correction for 3.6 m (12 ft) of 450 mm (18 in) duct	−2	−2	−1	−1	−1	−1	−1	−1
From Table 2.1, correction for 6.6 m (22 ft) of 350 mm (14 in) duct	−4	−4	−3	−2	−1	−1	−1	−1
From Table 2.1, correction for 9 m (30 ft) of 200 mm (8 in) duct	−6	−6	−4	−3	−2	−2	−2	−2
From Table 2.1, correction for 2.4 m (8 ft) of 100 mm (4 in) duct	−2	−2	−1	−1	−1	−1	−1	−1
From Table 2.2, correction for two 90° elbows at 350 mm (14 in)	0	0	−2	−16	−14	−8	−6	−6
From Table 2.2, correction for three 90° elbows at 200 mm (8 in)	0	0	0	−3	−24	−21	−12	−9
From Table 2.3, correction for 0.06 m^2 (100 in^2) outlet area	−12	−8	−3	−1	0	0	0	0
Total system attenuation =	−26	−22	−14	−27	−43	−34	−23	−20
PWL entering room (dB re 0.0002 dyne/cm^2) =	56	62	70	55	38	46	53	50

FIG 2.1 (Imperial) Ventilation system

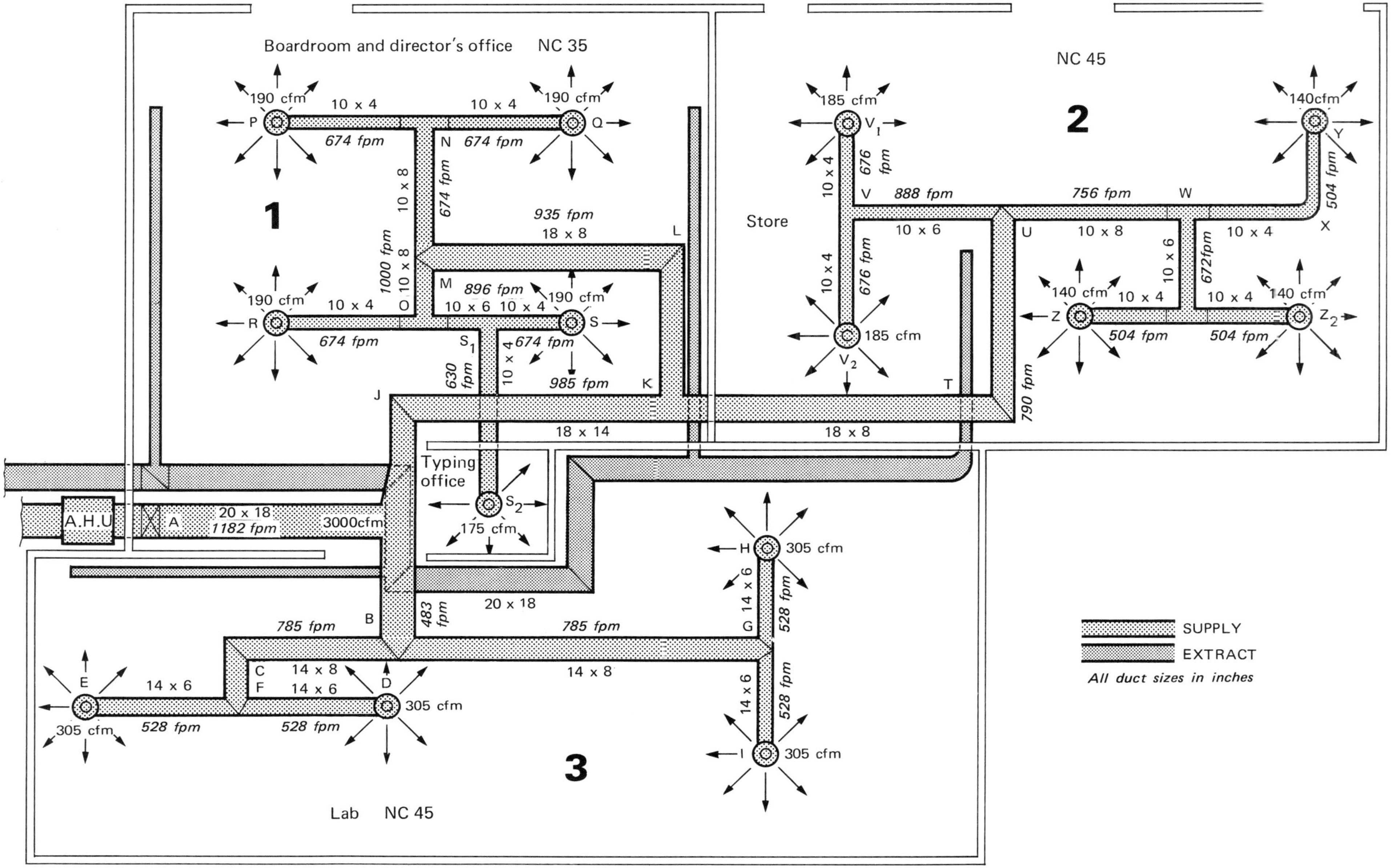

Boardroom and director's office NC 35

1

2

3

NC 45

Store

Typing office

Lab NC 45

SUPPLY

EXTRACT

All duct sizes in inches

Table 2.1 Attenuation of straight ducts (dB per 0.3m (per ft))

Straight duct (unlined)	D (mm)	D (in)	63 Hz	1250 Hz	250 Hz	500 Hz	1 kHz	2 kHz	4 kHz	8 kHz
Sheet steel	75– 180	3 – 7	0.2	0.2	0.15	0.1	0.1	0.1	0.1	0.1
	205– 380	8 –15	0.2	0.2	0.15	0.1	0.07	0.07	0.07	0.07
	405– 760	16 –30	0.2	0.2	0.1	0.05	0.05	0.05	0.05	0.05
	815–1525	32 –60	0.1	0.05	0.05	0.03	0.02	0.02	0.02	0.02
Sheet steel external lagging	75– 180	3 – 7	0.4	0.4	0.3	0.1	0.1	0.1	0.1	0.1
	205– 380	8 –15	0.4	0.4	0.3	0.1	0.07	0.07	0.07	0.07
	404– 760	16 –30	0.4	0.4	0.2	0.05	0.05	0.05	0.05	0.05
	815–1525	32 –00	0.1	0.1	0.1	0.03	0.02	0.02	0.02	0.02
Rigid walled	75– 180	3 – 7	0.03	0.03	0.05	0.05	0.1	0.1	0.1	0.1
	205– 380	8 –15	0.03	0.03	0.03	0.05	0.07	0.07	0.07	0.07
	405– 760	16 –30	0.02	0.02	0.02	0.03	0.05	0.05	0.05	0.05
	815–1525	32 –60	0.01	0.01	0.01	0.02	0.02	0.02	0.02	0.02
Rigid walled external lagging	76– 178	3 – 7	0.06	0.06	0.1	0.05	0.1	0.1	0.1	0.1
	203– 381	8 –15	0.06	0.06	0.06	0.05	0.07	0.07	0.07	0.07
	406– 762	16 –30	0.04	0.04	0.04	0.03	0.05	0.05	0.05	0.05
	813–1524	32 –60	0.02	0.02	0.02	0.02	0.02	0.02	0.02	0.02
Straight duct (lined)	**D (mm)**	**D (in)**	**63 Hz**	**125 Hz**	**250 Hz**	**500 Hz**	**1 kHz**	**2 kHz**	**4 kHz**	**8 kHz**
Sheet steel	75	3	0.2	0.2	0.8	3.6	8.5	12.0	12.0	12.0
	125	5	0.2	0.2	0.4	2.4	5.5	7.6	6.4	6.4
	205	8	0.2	0.2	0.2	1.6	3.8	4.6	2.0	1.0
	405	16	0.1	0.1	0.1	1.1	2.2	1.0	0.1	0.1
Sheet steel external lagging	75	3	0.4	0.4	1.6	3.6	8.5	12.0	12.0	12.0
	125	5	0.4	0.4	0.8	2.4	5.5	7.6	6.4	6.4
	205	8	0.4	0.4	0.4	1.6	3.8	4.6	2.0	1.0
	405	16	0.2	0.2	0.2	1.1	2.2	1.0	0.1	0.1
Rigid walled	75	3	0.2	0.2	6.8	3.6	8.5	12.0	12.0	12.0
	125	5	0.2	0.2	0.4	2.4	5.5	7.6	6.4	6.4
	205	8	0.2	0.2	0.2	1.6	3.8	4.6	2.0	1.0
	405	16	0.1	0.1	0.1	1.1	2.2	1.0	0.1	0.1
Rigid walled external lagging	75	3	0.4	0.4	1.6	3.6	8.5	12.0	12.0	12.0
	125	5	0.4	0.4	0.8	2.4	5.5	7.6	6.4	6.4
	205	8	0.4	0.4	0.4	1.6	3.8	4.6	2.0	1.0
	405	16	0.2	0.2	0.2	1.1	2.2	1.0	0.1	0.1

Unlined ducts provide a small amount of attenuation which increases by the application of a suitable lining material. The table gives the attenuation per foot (0.3 m) of various sizes and types of ducts. In all calculations the minimum dimension should be used to evaluate the attenuation.

Table 2.2 Attenuation of bends and elbows (dB)

Elbow and branches (unlined)	D (mm)	D (in)	63 Hz	1250 Hz	250 Hz	500 Hz	1 kHz	2 kHz	4 kHz	8 kHz
Radius sheet steel on rigid walled (D, R)	75– 140	3 – 5½	0	0	0	0	0	1	2	3
	150– 280	6 –11	0	0	0	0	1	2	3	3
	305– 585	12 –23	0	0	0	1	2	3	3	3
	610– 965	24 –33	0	1	1	2	3	3	3	3
Values apply if R is less than D/4	990–1980	39 –78	1	2	2	3	3	3	3	3
Vaned sheet steel or rigid walled (D)	75– 140	3 – 5¼	0	0	0	0	0	1	2	3
	150– 280	6 –11	0	0	0	0	2	3	3	3
	305– 585	12 –23	0	0	0	2	3	4	3	3
	610– 965	24 –33	0	2	2	3	4	3	3	3
	990–1980	39 –78	2	3	3	4	3	3	3	3
Square sheet steel or rigid walled (D)	75– 100	3 – 4	0	0	0	0	1	8	6	4
	115– 140	4½– 5½	0	0	0	0	4	7	5	3
	150– 205	6 – 8	0	0	0	1	8	7	4	3
	230– 280	9 –11	0	0	0	4	7	5	3	3
	305– 405	12 –16	0	0	1	8	7	4	3	3
	430– 585	17 –23	0	0	4	7	5	4	3	3
	610– 840	24 –33	0	1	8	7	4	3	3	3
	865– 965	34 –33	0	4	7	5	4	3	3	3
	990–1980	39 –78	0	4	7	5	3	3	3	3

Elbows and branches (lined)	D (mm)	D (in)	63 Hz	1250 Hz	250 Hz	500 Hz	1 kHz	2 kHz	4 kHz	8 kHz
Square full lined (D, t)	76– 102	3 – 4	0	0	0	0	2	13	18	18
	114– 140	4½– 5½	0	0	0	0	7	16	18	16
	152– 203	6 – 8	0	0	0	2	13	18	18	16
	229– 279	9 –11	0	0	0	7	16	18	16	17
	305– 406	12 –16	0	0	2	13	18	18	16	18
	432– 584	17 –23	0	0	7	16	18	16	17	18
	610– 833	24 –33	0	2	13	18	18	16	18	17
Thickness t = approx. 10%D	864– 965	34 –33	0	7	16	18	16	17	18	16
Square half-lined (D, t)	75– 100	3 – 4	0	0	0	0	2	11	13	12
	115– 140	4½– 5½	0	0	0	0	6	14	13	11
	150– 205	6 – 8	0	0	0	2	11	13	12	10
	230– 280	9 –11	0	0	0	6	14	13	11	10
	305– 405	12 –16	0	0	2	11	13	12	10	10
	430– 585	17 –23	0	0	6	14	13	11	10	10
	610– 840	24 –33	0	2	11	13	12	10	10	10
Thickness t = approx. 10%D	865– 965	34 –38	0	2	14	13	11	10	10	10

Bends generally provide relatively poor sound reduction at the lower frequencies.
Lining of suitable absorption material, placed just after the elbow where repeated reflection takes place, improves the attenuation. The table gives the values of attenuation for various types of elbows.

Table 2.3 Attenuation of outlets (dB) (Due to end reflection loss)

The attenuation of outlets is almost negligible for high frequency sound, i.e. all the power in the high frequencies bands is allowed to pass through to the room.
For low frequencies, however, some of it is reflected back. Thus substantial low frequency attenuation is achieved. Attenuation is also dependent upon the total area of the outlet.

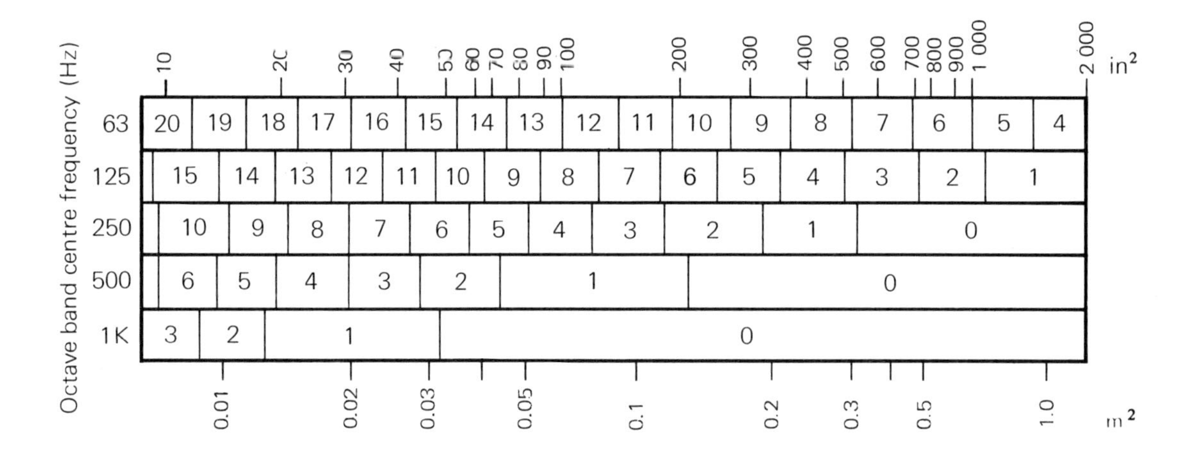

TABLE 2.4 Percentage correction (dB)

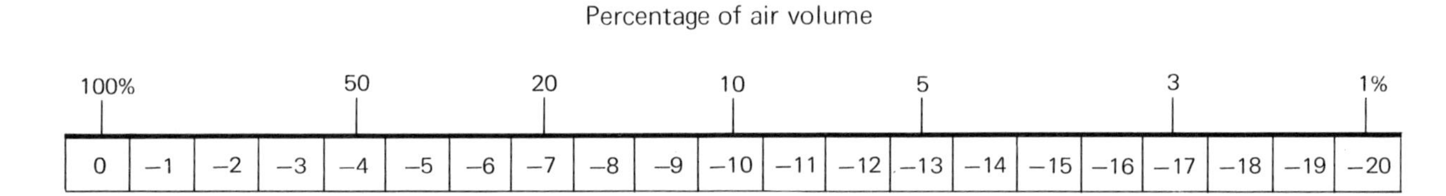

The sound power level emerging from outlets into a room is in approximately the same proportion as the airflow from these outlet. Therefore, initially estimate the percentage of air entering the room. The sound reduction based on this percentage can then be read from the table.

TABLE 2.5 Reverberation time and room volume correction tables

(a) Correction for reverberation time

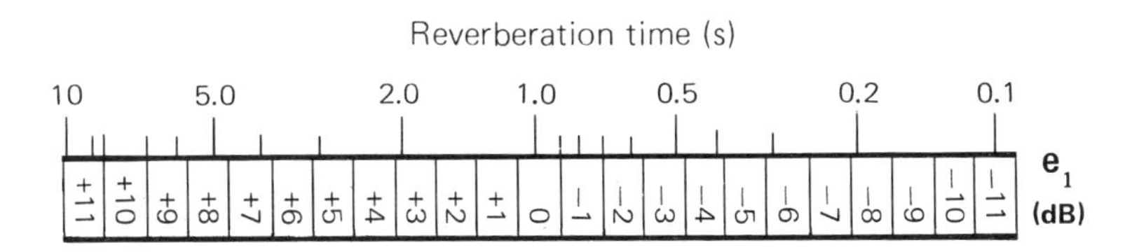

(b) Correction for room volume

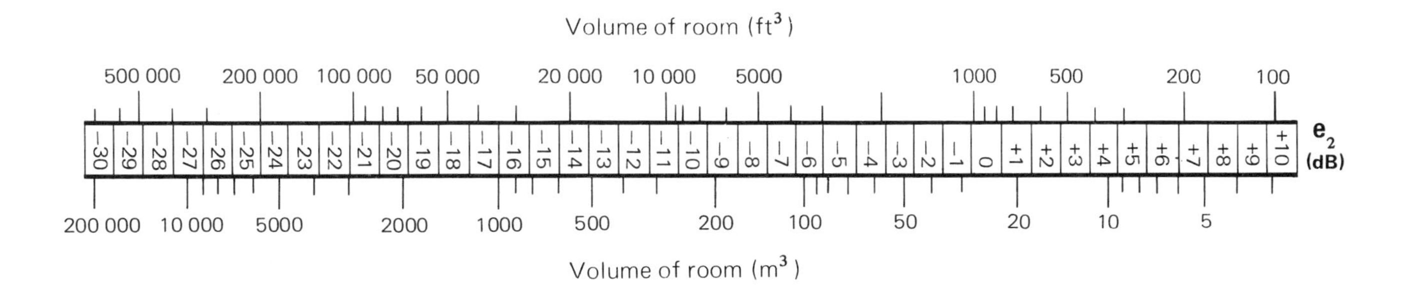

(c) Reverberation times for various types of rooms

Use of room	*Reverberation time (s)*	*Rating*
Radio and TV studios, music rooms	0.2 to 0.25	Dead
Department stores, restaurants	0.4 to 0.5	Medium dead
Offices, libraries, homes, hotel rooms, conference rooms, lecture halls	0.9 to 1.1	Average
School rooms, art galleries, building lobbies, hospitals, small churches	1.8 to 2.2	Medium live
Large churches, gymnasiums, factories	2.5 to 4.5	Live

2.4.2 Reverberant sound pressure level

The reverberant sound pressure level in a room is dependent upon the room characteristics, e.g. whether the room is 'alive' or 'dead'. The simplest way of defining room characteristics is by the reverberation time and volume, although an alternative way is by surface area and average absorption coefficient of the surfaces. The latter, however, involves more work. To calculate reverberant sound pressure level, initially one should estimate the proportion of sound power entering the room as a percentage of the total fan sound power. For low velocity systems this is (approximately) in the same proportion as the airflow emerging from the outlets in that room. For instance, if the room is receiving 10 per cent of the total airflow, then the sound power entering the room will be 10 per cent of the total.

Factors for various percentages are given in Table 2.4. Thus if e_1, e_2 and d are the factors read from Table 2.5 (a) and (b) and Table 2.4:

$$\text{Reverberant SPL} = \text{power level entering room} - (d + e_1 + e_2).$$

A correction must be added to this if the room is served by one or more additional fans. Most rooms are served by at least one additional fan for the extract duct system. The factor for one additional fan is +3 dB. In some cases, a room may be served by more than two separate fans, and the factors for these are given in Table 2.8.

EXAMPLE 3:

Calculate the reverberant sound pressure level in a room of 140 m^3 (5000 ft^3) volume, and getting 25% of the total air. The sound power level leaving the system is given at the end of Example 2. The reverberation time is 1½ seconds.

SOLUTION

	63	125	250	500	1000	2000	4000	8000 Hz
PWL entering room	56	62	70	55	38	46	53	50
From Table 2.5(a), correction for 1½ s rev. time	+2	+2	+2	+2	+2	+2	+2	+2
From Table 2.4, correction for 25 per cent air	−6	−6	−6	−6	−6	−6	−6	−6
From Table 2.5(b), correction for 140 m^3 (5000 ft^3) volume	−8	−8	−8	−8	−8	−8	−8	−8
Total correction	−12	−12	−12	−12	−12	−12	−12	−12
Reverberant SPL	44	50	58	43	26	34	41	38
From Table 2.8, +3 dB for extract system	47	53	61	46	29	37	44	41

2.4.3 Direct sound pressure level

Direct sound pressure level is due to the presence of a noise source (e.g. grille) near the listener's ear. The estimation is, therefore, based on whichever outlet is considered nearest to the listener's ear.

The relevant parameters are:

(1) The percentage of air handled by that outlet *only*.
(2) The distance of that outlet from the ear.
(3) The direction or directivity of the noise emission from the outlet.

The first factor can be evaluated from Table 2.4 as in the case of the reverberant level.

TABLE 2.6 Distance and directivity correction tables

(a) Correction for distance

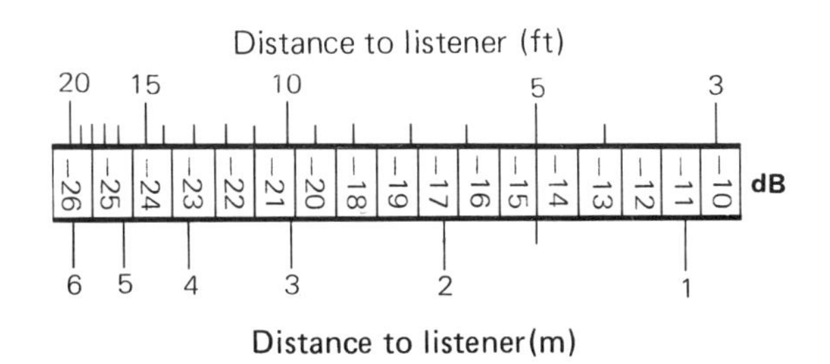

In most common applications such as offices, schools and hotel bedrooms the distance may be chosen as 1.5m (5 ft).

(b) Correction for directivity

Flush with surface

+3 +4 +5 +6 +7 63 Hz
+3 +4 +5 +6 +7 +8 125
+3 +4 +5 +6 +7 +8 250
+3 +4 +5 +6 +7 +8 +9 500
+4 +5 +6 +7 +8 +9 1000
+5 +6 +7 +8 +9 2000
+7 +8 +9 4000
+8 +9 8000
1 10 100 1000 10 000
Outlet area (in^2)
0.001 0.01 0.1 1 10
Outlet area (m^2)

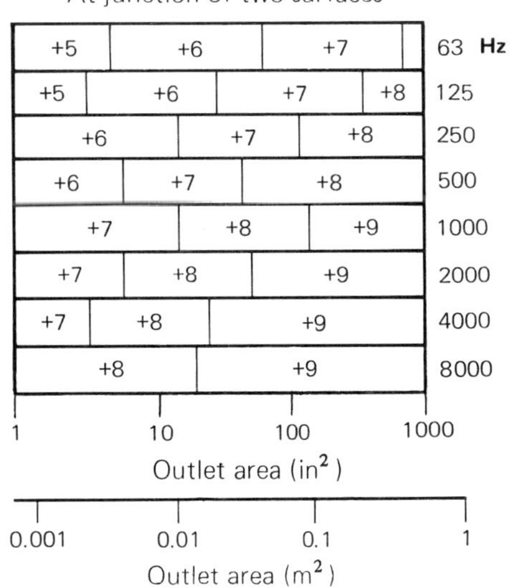

At junction of three surfaces
(ceiling and two walls)

+9 63 Hz
+9 125
+9 250
+9 500
+9 1000
+9 2000
+9 4000
+9 8000
Any outlet area

The factors for distance and directivity are given in the Tables 2.6(a) and 2.6(b). For most common applications such as offices, schools and hotel bedrooms the distance may be taken as 1.5 m (5 ft). A correction must also be made for additional fan systems serving the room (Table 2.8).

EXAMPLE 4:

Calculate the direct sound pressure level due to an air outlet of 0.06 m^2 (100 in^2) flush with the ceiling and handling 6 per cent of the total air.

SOLUTION

	63	125	250	500	1000	2000	4000	8000 Hz
Sound power level leaving system	56	62	70	55	38	46	53	50
From Table 2.6(a), correction for 1.5 m (5 ft) distance to ear	−14	−14	−14	−14	−14	−14	−14	−14
From Table 2.6(b), correction for directivity 0.06 m^2 (100 in^2) diffuser flush with surface	+4	+4	+5	+7	+8	+8	+9	+9
From Table 2.4, correction for 6 per cent air	−12	−12	−12	−12	−12	−12	−12	−12
From Table 2.8, +3 dB for nearby extract grille	+3	+3	+3	+3	+3	+3	+3	+3
Total attenuation	−19	−19	−18	−16	−15	−15	−14	−14
Direct SPL	37	43	52	39	23	31	39	36

2.4.4 Resultant sound pressure level

Resultant sound pressure level is the logarithmic sum of reverberant and direct sound pressure levels. This can be found using Table 2.7.

TABLE 2.7 Addition of decibels

Difference between SPLs	*Add to larger SPL*
0,1	+3
2,3	+2
4,5,6,7,8,9	+1
10 or more	0

TABLE 2.8 Correction for additional fans

Number of additional fans serving room	*Add*
1	+3 dB
2	+5 dB
3	+6 dB
4	+7 dB

EXAMPLE 5:

Calculate the resultant room sound pressure level for the critical room from the previous examples.

SOLUTION

	63	125	250	500	1000	2000	4000	8000	Hz
Reverberant sound pressure level	47	53	61	46	29	37	44	41	
Direct sound pressure level	37	43	52	39	23	31	39	36	
From Table 2.7, combined SPL	47	53	62	47	30	38	45	42	

2.5 Establishing the noise design criterion.

This is an important decision, and too often the noise criterion is plucked from one of the guide books without sufficient thought being given to what such a criterion would mean in terms of additional cost or whether it really will be the most suitable for the application. As an example NC 30 may be a suitable criterion for a boardroom, if the building is in a quiet rural area, or if the room itself is in the centre core surrounded by other quiet areas. On the other hand, if it has single glazed windows looking out on to busy streets, the noise from other sources such as traffic may result in an environment of NC 40. It must be remembered that the principle aim is not to silence the mechanical services, but merely to make them inaudible against the background level. There are other examples where the ventilation system noise is purposely chosen at a higher level in order to mask intrusive noises and provide speech privacy.

In view of the importance of this criterion, it is advisable to discuss it with the architect. Desirable NC levels for some common environments are shown in Table 2.9.

TABLE 2.9 Acceptable noise criteria

	NC values
Broadcasting and recording studios	20
Concert halls and theatres	20-25
Assembly halls and churches	25-30
Cinemas	30-35
Hospital wards and operating theatres	30-40
Homes, bedrooms	25-35
Private offices and libraries	30-35
General offices	35-45
Mechanised offices	40-55
Restaurants, bars	35-45
Department stores	35-45
Swimming baths and sports arenas	35-50

The ventilation system noise should not exceed the levels shown here.

2.6 Establishing the insertion loss of a suitable silencer.

An NC number represents a certain sound pressure level in each octave band. This pressure level is then compared with the resulting sound pressure level in the room, calculated by the procedure described above. The number of decibels by which the resultant pressure level exceeds the NC levels in the room in each frequency band, gives the insertion loss of a silencer required to bring the resultant pressure level within the NC limits (Table 2.10).

TABLE 2.10 Sound pressure levels corresponding to various NC levels

NC levels	*Octave band (Hz)*							
	63	125	250	500	1000	2000	4000	8000
65	80	75	71	68	66	64	63	62
60	77	71	67	63	61	59	58	57
55	74	67	62	58	56	54	53	52
50	71	64	58	54	51	49	48	47
45	67	60	54	49	46	44	43	42
40	64	57	50	45	41	39	38	37
35	60	52	45	40	36	34	33	32
30	57	48	41	35	31	29	28	27
25	54	44	37	31	27	24	22	21

EXAMPLE 6:

Calculate the additional silencing for the layout in Fig. 2.1 to meet the requirements of NC 35.

SOLUTION

	63	125	250	500	1000	2000	4000	8000	Hz
Resultant room SPL	47	53	62	47	30	38	45	42	
NC 35	60	52	45	40	36	34	33	32	
Required silencing	0	1	17	7	0	4	12	10	

2.7 Silencer selection.

Appendix 3 gives typical silencer data. Its use can best be illustrated by continuing Example 6.

For this application:

Duct size = 0.225 m^2 (2.5 ft^2)
Airflow = 5100 m^3/h (3000 ft^3/min)
Velocity = 6.0 m/s (1200 ft/min) approximately
Insertion loss required = 17 dB in 250 Hz band.

From Appendix 3 a silencer 1.2 m (48 in) long gives the required silencing.

Suitable dimensions for inserting into 0.45 m x 0.5 m (18 in x 20 in) duct would be 0.45 m x 0.7 m (18 in x 28 in), using an expansion and contraction piece, giving a net free area = 0.14 m^2 (1.5 ft^2) (see note 7 on p.4).

Passage velocity therefore $= \dfrac{5100\ m^3/h}{0.14\ m^2} = 36000\ m/h = 10\ m/s \left(\dfrac{3000\ ft^3/min}{1.5\ ft^2} = 2000\ ft/min\right)$

Air pressure drop = 50 Pascal (0.2 in wg).

Velocity generated noise at 10 m/s (2000 ft/min) = 33 dB (Fig 3.8) which is well below the existing sound power level of 84 dB.

2.8 Extract systems.

It is important to remember that extract systems also require acoustic treatment. When air is extracted by a separate duct system, the extract ductwork must be analysed in a manner similar to the supply system.

Where ceiling voids are utilised to extract air, they may be treated like small rooms from which noise can break out into the conditioned space.

2.9 Breakout noise.

The breakout of noise through the duct walls adds to the noise level in a room. In many cases the breakout noise is greater than that transmitted via the grilles.

The breakout noise calculations in section 4.2 indicate how this problem is tackled, and show how the breakout noise adds to the noise transmitted via the grilles, thus increasing the silencing requirement.

In this instance the silencing for the 250 Hz band will probably be the most critical, i.e. if this band is covered the other octave bands will automatically be silenced.

2.10 Calculation method.

The method of carrying out calculations based on the procedure discussed above is described in many publications. One such publication is *Design for Sound*, published by Woods of Colchester Ltd. Caution should be exercised, however, in the use of these guides, as most of them do not give the data for the 63 Hz band. This band in many cases becomes critical and should be included in the calculations.

EXAMPLE 7:

For the duct system layout shown in Fig. 2.1 carry out a complete ductborne noise analysis and determine the size of the main silencer needed to meet the acoustic requirements of all the rooms shown. The fan is 0.4m (15 in) in diameter and the power level spectrum is as follows:—

Frequency	63	125	250	500	1000	2000	4000	8000	Hz
Fan sound power level (dB re 0.0002 dyne/cm^2)	82	84	84	82	81	80	76	70	

SOLUTION

It has already been shown in the example on page 12 that the critical duct branches are ABC and AJKL. These branches are analysed in the charts in Figs. 2.2 and 2.3.

SILENCER SIZES

The analysis shows that a silencer capable of giving 15 dB attenuation in the 250 Hz band would satisfy all the requirements.

COMPUTER PROGRAM

A simple basic time sharing computer program is available to do this type of calculation in all the eight octave bands from 63 to 8000 Hz. Details are given in Appendix 5.

FIG. 2.2 (metric) Ventilation noise analysis chart I

Job title *(Duct run A. BC)* Supply to Laboratory	*Frequency (Hz)* 63	125	250	500	1000	2000	4000	8000	Table
a Fan PWL = PWL entering system	82	84	84	82	81	80	76	70	
b Bend/duct — Size (mm) — Length/angle — Treatment									
Duct — 450 — 3·6 m — none	$-1\frac{1}{2}$	$-1\frac{1}{2}$	−1	$-\frac{1}{2}$	−1	−1	−1	−1	2.1
Elbow — 350 — 90° — none	0	0	−1	−8	−7	−4	−3	−3	2.2
Duct — 350 — 1·8 m — none	−1	−1	−1	$-\frac{1}{2}$	$-\frac{1}{2}$	$-\frac{1}{2}$	$-\frac{1}{2}$	$-1\frac{1}{2}$	2.1
B Elbow — 200 — 90° — none	0	0	0	−1	−8	−7	−4	−3	2.2
Duct — 200 — 3·0 m — none	$-1\frac{1}{2}$	$-1\frac{1}{2}$	−1	−1	$-\frac{1}{2}$	$-\frac{1}{2}$	$-\frac{1}{2}$	$-\frac{1}{2}$	2.1
C Elbow — 200 — 90° — none	0	0	0	−1	−8	−7	−4	−3	2.2
F Elbow — 150 — 90° — none	0	0	0	−1	−8	−7	−4	−3	2.2
Duct — 150 — 2·4 m — none	$-1\frac{1}{2}$	$-1\frac{1}{2}$	−1	−1	−1	−1	−1	−1	2.1
Single outlet area = 0·09 m²	−12	−7	−3	−1	0	0	0	0	2.3
c Total system attenuation (Σb)	17	12	8	15	34	28	18	16	
d PWL leaving system ($a-c$)	65	72	76	67	47	52	58	54	
e Percentage of sound reaching room = 40%	−4	−4	−4	−4	−4	−4	−4	−4	2.4
f Room volume = 308 m³	−11	−11	−11	−11	−11	−11	−11	−11	2.5(b)
g Reverb. time = 1·7 s	+2	+2	+2	+2	+2	+2	+2	+2	2.5(a)
h Factor for additional fans serving the room (extract, etc.)	+3	+3	+3	+3	+3	+3	+3	+3	2.8
i Reverberation SPL correction ($e+f+g+h$)	−10	−10	−10	−10	−10	−10	−10	−10	
j Reverberant SPL ($d-i$)	55	62	66	57	37	42	48	44	
k Percentage of sound reaching output = 10%	−10	−10	−10	−10	−10	−10	−10	−10	2.4
l Distance to listener = 1·5 m	−14	−14	−14	−14	−14	−14	−14	−14	2.6(a)
m Directivity (outlet area) = 0·09 m²	+5	+6	+7	+8	+8	+9	+9	+9	2.6(b)
Factor for additional fans serving the room (extract, etc.)	+3	+3	+3	+3	+3	+3	+3	+3	2.8
n Direct SPL correction ($k+l+m+h$)	−16	−15	−14	−13	−13	−12	−12	−12	
o Direct SPL ($d-h$)	49	57	62	54	34	40	46	42	
p Resultant room SPL	56	63	67	59	39	44	50	46	2.7
q NC requirement 45	67	60	54	49	46	44	43	42	2.10
w_1 Treatment ($p-q$) if $p>q$	0	3	13	10	0	0	7	4	
r Duct breakout noise for duct over critical area *s* T.L. of ceiling *t* Breakout power level entering room ($r-s$) *u* Correction for room volume and characteristic									
v Room SPL due to duct breakout noise ($t-u$)									
q NC requirement									
w_2 Additional treatment ($v-q$) if $v>q$									
w_1 or w_2 whichever is greater									

FIG. 2.2 (Imperial) Ventilation noise analysis chart I

	Job title (Duct run A, BC) Supply to Laboratory				63	125	250	500	1000	2000	4000	8000	Table
					Frequency (Hz)								
a	Fan PWL = PWL entering system				82	84	84	82	81	80	76	70	
b	Bend/duct	Size (in)	Length/angle	Treatment									
	Duct	18	12 ft	none	-1½	-1½	-1	-½	-1	-1	-1	-1	2.1
	Elbow	14	90°	none	0	0	-1	-8	-7	-4	-3	-3	2.2
	Duct	14	6 ft	none	-1	-1	-1	-½	-½	-½	-½	-1½	2.1
	B Elbow	8	90°	none	0	0	0	-1	-8	-7	-4	-3	2.2
	Duct	8	10 ft	none	-1½	-1½	-1	-1	-½	-½	-½	-½	2.1
	C Elbow	8	90°	none	0	0	0	-1	-8	-7	-4	-3	2.2
	F Elbow	6	90°	none	0	0	0	-1	-8	-7	-4	-3	2.2
	Duct	6	8 ft	none	-1½	-1½	-1	-1	-1	-1	-1	-1	2.1
	Single outlet area = 144 in²				-12	-7	-3	-1	0	0	0	0	2.3
c	Total system attenuation (Σb)				17	12	8	15	34	28	18	16	
d	PWL leaving system ($a-c$)				65	72	76	67	47	52	58	54	
e	Percentage of sound reaching room = 40%				-4	-4	-4	-4	-4	-4	-4	-4	2.4
f	Room volume = 11000 ft³				-11	-11	-11	-11	-11	-11	-11	-11	2.5(b)
g	Reverb. time = 1·7 s				+2	+2	+2	+2	+2	+2	+2	+2	2.5(a)
h	Factor for additional fans serving the room (extract, etc.)				+3	+3	+3	+3	+3	+3	+3	+3	2.8
i	Reverberation SPL correction ($e+f+g+h$)				-10	-10	-10	-10	-10	-10	-10	-10	
j	Reverberant SPL ($d-i$)				55	62	66	57	37	42	48	44	
k	Percentage of sound reaching outlet = 10%				-10	-10	-10	-10	-10	-10	-10	-10	2.4
l	Distance to listener = 5 ft				-14	-14	-14	-14	-14	-14	-14	-14	2.6(a)
m	Directivity (outlet area) = 144 in²				+5	+6	+7	+8	+8	+9	+9	+9	2.6(b)
	Factor for additional fans serving the room (extract, etc.)				+3	+3	+3	+3	+3	+3	+3	+3	2.8
n	Direct SPL correction ($k+l+m+h$)				-16	-15	-14	-13	-13	-12	-12	-12	
o	Direct SPL ($d-n$)				49	57	62	54	34	40	46	42	
p	Resultant room SPL				56	63	67	59	39	44	50	46	2.7
q	NC requirement 45				67	60	54	49	46	44	43	42	2.10
w_1	Treatment ($p-q$) if $p>q$				0	3	13	10	0	0	7	4	
r	Duct breakout noise for duct over critical area												
s	T.L. of ceiling												
t	Breakout power level entering room ($r-s$)												
u	Correction for room volume and characteristic												
v	Room SPL due to duct breakout noise ($t-u$)												
q	NC requirement												
w_2	Additional treatment ($v-q$) if $v>q$												
	w_1 or w_2 whichever is greater												

FIG. 2.3 (metric) Ventilation noise analysis chart II

Job title (*Duct run AJKL*)	*Frequency (Hz)*								
Supply to Boardroom and Director's Suite	63	125	250	500	1000	2000	4000	8000	Table
a Fan PWL = PWL entering system	82	84	84	82	81	80	76	70	
b Bend/duct — Size (mm) — Length/angle — Treatment									
Duct — 450 — 3.6 m — none	$-2\frac{1}{2}$	$-2\frac{1}{2}$	−1	−1	−1	−1	−1	−1	2.1
2 square elbows — 350 — 90° — none	0	0	−2	−16	−14	−8	−6	−6	2.2
Duct — 350 — 6.6 m — none	$-4\frac{1}{2}$	$-4\frac{1}{2}$	−3	−2	−1	−1	−1	−1	2.1
3 square elbows — 200 — 90° — none	0	0	0	−3	−24	−21	−12	−9	2.2
Duct — 200 — 9 m — none	−6	−6	−4	−3	−2	−2	−2	−2	2.1
Duct — 100 — 2.4 m — none	$-1\frac{1}{2}$	$-1\frac{1}{2}$	−1	−1	−1	−1	−1	−1	2.1
Single outlet area = 0.06 m^2	−12	−8	−4	−1	0	0	0	0	2.3
c Total system attenuation (Σb)	−26	−22	−15	−27	−43	−34	−23	−20	
d PWL leaving system (*a*–*c*)	56	62	69	55	38	46	53	50	
e Percentage of sound reaching room = 25%	−6	−6	−6	−6	−6	−6	−6	−6	2.4
f Room volume = 140 m^3	−8	−8	−8	−8	−8	−8	−8	−8	2.5(b)
g Reverb. time = $1\frac{1}{2}$ s	+2	+2	+2	+2	+2	+2	+2	+2	2.5(a)
h Factor for additional fans serving the room (extract, etc.)	+3	+3	+3	+3	+3	+3	+3	+3	2.8
i Reverberation SPL correction ($e+f+g+h$)	−9	−9	−9	−9	−9	−9	−9	−9	
j Reverberant SPL (*d*–*i*)	47	53	60	46	29	37	44	41	
k Percentage of sound reaching outlet = 6%	−12	−12	−12	−12	−12	−12	−12	−12	2.4
l Distance to listener = 1.5 m	−14	−14	−14	−14	−14	−14	−14	−14	2.6(a)
m Directivity (outlet area) = 0.06 m^2	+4	+4	+5	+7	+8	+8	+9	+9	2.6(b)
Factor for additional fans serving the room (extract, etc.)	+3	+3	+3	+3	+3	+3	+3	+3	2.8
n Direct SPL correction ($k+l+m+h$)	−19	−19	−18	−16	−15	−15	−14	−14	
o Direct SPL (*d*–*n*)	37	43	51	39	23	31	39	36	
p Resultant room SPL	47	53	61	47	30	38	45	42	2.7
q NC requirement 35	60	52	45	40	36	34	33	32	2.10
w_1 Treatment (p–q) if $p>q$	0	1	16	7	0	4	12	10	
r Duct breakout noise for duct over critical area *s* T.L. of ceiling *t* Breakout power level entering room (*r*–*s*) *u* Correction for room volume and characteristic									
v Room SPL due to duct breakout noise (*t*–*u*)									
q NC requirement									
w_2 Additional treatment (v–q) if $v>q$ w_1 or w_2 whichever is greater									

FIG. 2.3 (Imperial) Ventilation noise analysis chart II

Job title (*Duct run A.JKL*)				*Frequency (Hz)*								
Supply to Boardroom and Director's Suite				63	125	250	500	1000	2000	4000	8000	Table
a Fan PWL = PWL entering system				82	84	84	82	81	80	76	70	
b Bend/duct	Size (in)	Length/angle	Treatment									
Duct	18	12 ft	none	$-2\frac{1}{2}$	$-2\frac{1}{2}$	−1	−1	−1	−1	−1	−1	2.1
2 square elbows	14	90°	none	0	0	−2	−16	−14	−8	−6	−6	2.2
Duct	14	22 ft	none	$-4\frac{1}{2}$	$-4\frac{1}{2}$	−3	−2	−1	−1	−1	−1	2.1
3 square elbows	8	90°	none	0	0	0	−3	−24	−21	−12	−9	2.2
Duct	8	30 ft	none	−6	−6	−4	−3	−2	−2	−2	−2	2.1
Duct	4	8 ft	none	$-1\frac{1}{2}$	$-1\frac{1}{2}$	−1	−1	−1	−1	−1	−1	2.1
Single outlet area = 100 in^2				−12	−8	−4	−1	0	0	0	0	2.3
c Total system attenuation (Σb)				−26	−22	−15	−27	−43	−34	−23	−20	
d PWL leaving system ($a-c$)				56	62	69	55	38	46	53	50	
e Percentage of sound reaching room = 25%				−6	−6	−6	−6	−6	−6	−6	−6	2.4
f Room volume = 5000 ft^3				−8	−8	−8	−8	−8	−8	−8	−8	2.5(b)
g Reverb. time = $1\frac{1}{2}$ s				+2	+2	+2	+2	+2	+2	+2	+2	2.5(a)
h Factor for additional fans serving the room (extract, etc.)				+3	+3	+3	+3	+3	+3	+3	+3	2.8
i Reverberation SPL correction ($e+f+g+h$)				−9	−9	−9	−9	−9	−9	−9	−9	
j Reverberant SPL ($d-i$)				47	53	60	46	29	37	44	41	
k Percentage of sound reaching outlet = 6%				−12	−12	−12	−12	−12	−12	−12	−12	2.4
l Distance to listener = 5 ft				−14	−14	−14	−14	−14	−14	−14	−14	2.6(a)
m Directivity (outlet area) = 100 in^2				+4	+4	+5	+7	+8	+8	+9	+9	2.6(b)
Factor for additional fans serving the room (extract, etc.)				+3	+3	+3	+3	+3	+3	+3	+3	2.8
n Direct SPL correction ($k+l+m+h$)				−19	−19	−18	−16	−15	−15	−14	−14	
o Direct SPL ($d-n$)				37	43	51	39	23	31	39	36	
p Resultant room SPL				47	53	61	47	30	38	45	42	2.7
q NC requirement 35				60	52	45	40	36	34	33	32	2.10
w_1 Treatment ($p-q$) if $p>q$				0	1	16	7	0	4	12	10	
r Breakout power level entering room ($r-s$) *s* T.L. of ceiling *t* Breakout power level entering room ($r-s$) *u* Correction for room volume and characteristic												
v Room SPL due to duct breakout noise ($t-u$)												
q NC requirement												
w_2 Additional treatment ($v-q$) if $v>q$												
w_1 or w_2 whichever is greater												

CHAPTER THREE

Velocity generated noise

3.1 Introduction.

The acoustic analysis discussed so far is basically a static noise analysis. Acoustically speaking the fan might just as well have been a loudspeaker (emitting sound of an identical power level spectrum) and the method described on the previous pages would provide a suitable remedy.

However, with the presence of air flowing in the ductwork, additional noise may be generated where the airflow meets obstructions, sharp bends, sudden enlargements or contractions. Even straight duct lengths give rise to this noise under certain conditions.

Silencers, like any other duct component, also generate noise at the higher air velocities. It is because of this that improperly designed installations can sometimes be noisier with silencers than without.

Velocity generated noise, also known as regenerated noise or self noise, is usually negligible for velocities up to 10 m/s (2000 ft/min) in the main risers. This maximum limit should be brought down to 7.5 m/s (1500 ft/min) for branch ducts and risers, and 4 m/s (800 ft/min) for ducts serving the grilles. Above these limits this noise will become a substantial component of the system noise.

3.2 Dealing with velocity generated noise.

For air velocities higher than those stated above it is essential that regenerated noise levels are evaluated and checked against static power levels.

Levels of velocity generated noise for most common types of duct fittings can be estimated from the data given in Figs. 3.2 to 3.8. For some duct components, this data refers only to specific sizes. For others however, it is presented in a generalised form. In this form (Figs. 3.2 to 3.4) the data is usually expressed as a function of the dimensionless Strouhal Number, which is a function of the size of the fitting, the velocity of the air and the frequency. For generalised data, therefore, the calculation of the Strouhal Number is the starting point.

$$\text{Strouhal Number } (N_{STR}) = \frac{fd}{v}$$

where d is the characteristic dimension
v is the air velocity
f is the band frequency
} in compatible units

These quantities must be in compatible units, for example m, m/s and hertz, but *not* mm, m/s and hertz.

The nomograph (Fig. 3.1) may be used as a convenient way of calculating the Strouhal Number. Place a straight edge joining the scales of duct diameter and air velocity and mark off a point on the reference line. From this reference point place the straight edge across to any chosen band frequency to read off the corresponding Strouhal Number.

From the Strouhal Numbers, determine the corresponding 'F' factors for the appropriate type of fitting. Add this to the appropriate 'G' and 'H' factors (which allow for air velocity and noise frequency respectively) to get the octave band sound power levels of the velocity generated noise, ie:–

$$\text{Octave band sound power level} = F + G + H$$

The levels arrived at in this way should then be compared with the power level existing at the room side end of that fitting and should always be at least 7 dB below it. If this is not so, then it must be acknowledged that the noise generated by that fitting will contribute to the overall system noise, and adequate steps (such as reducing the air velocity and/or smoothing the airflow path) must be taken.

In the calculation it is recommended that the velocity generated noise from the tables is evaluated and compared with power levels existing in each duct component. In this way, the problem areas can be identified where the difference between the two is narrower than 7 dB.

It must also be remembered that silencers, like any other duct component, give rise to velocity generated noise with higher air velocities. This noise also increases with the number of modules. It is because of this velocity generated noise that improperly designed installations can be noisier with silencers than without. Figure 3.8 gives the velocity generated sound power levels for a typical silencer.

3.3 Computer programs for velocity generated noise

Extensive calculations are required in order to arrive at the velocity generated noise power level (PWL) and these can be very laborious if a lot of them have to be performed manually.

Computer programs are available to predict the velocity generated noise of bends and branches based on the normalised data. Full details are given in Appendix 5.

FIG 3.1 Nomograph for calculating Strouhal Number

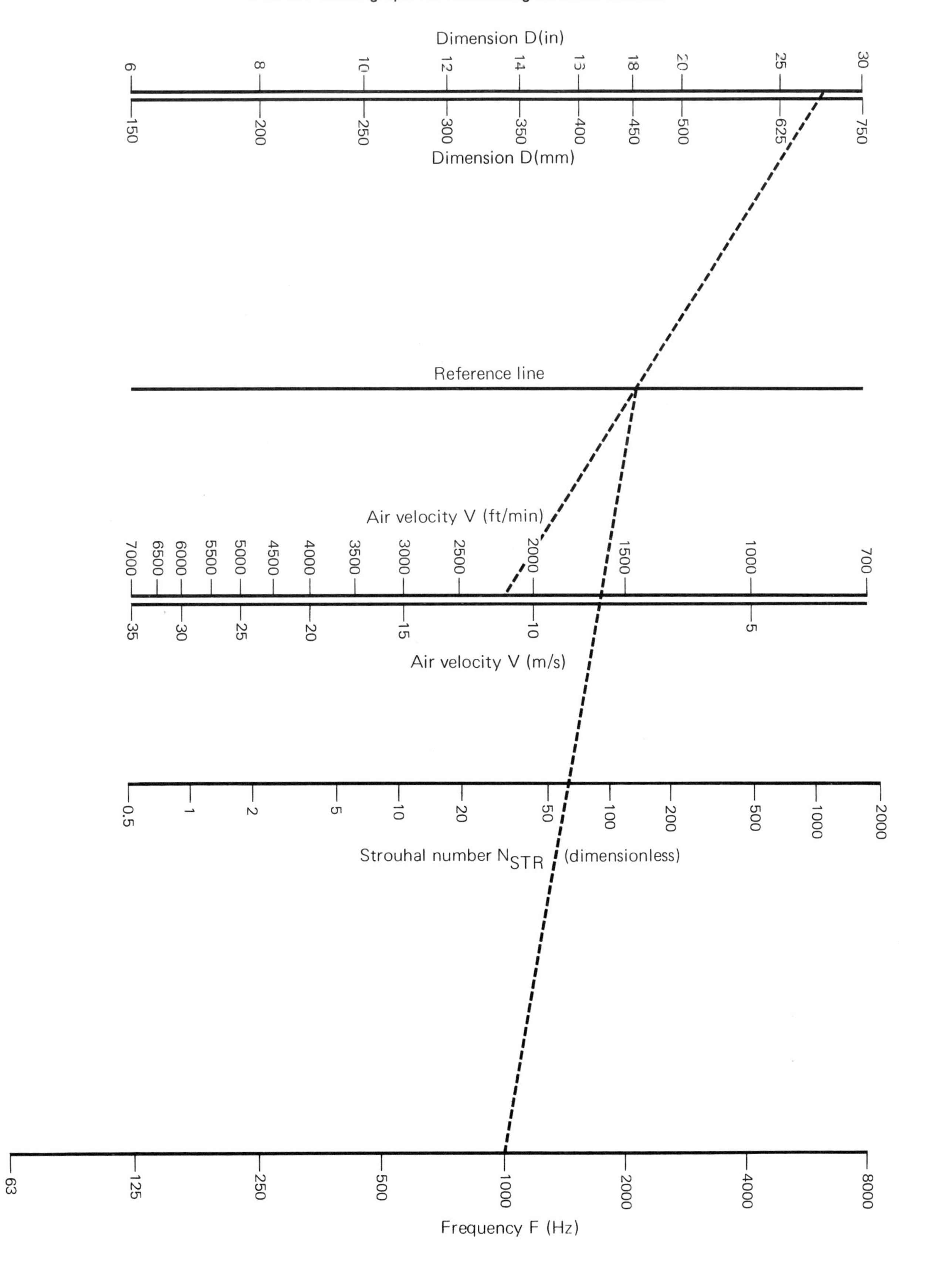

Based on data presented in ASHRAE Handbook (1973 Systems Volume ch 35)

FIG 3.2(a) 'F' factors for calculating velocity generated noise in duct elbows (bends)

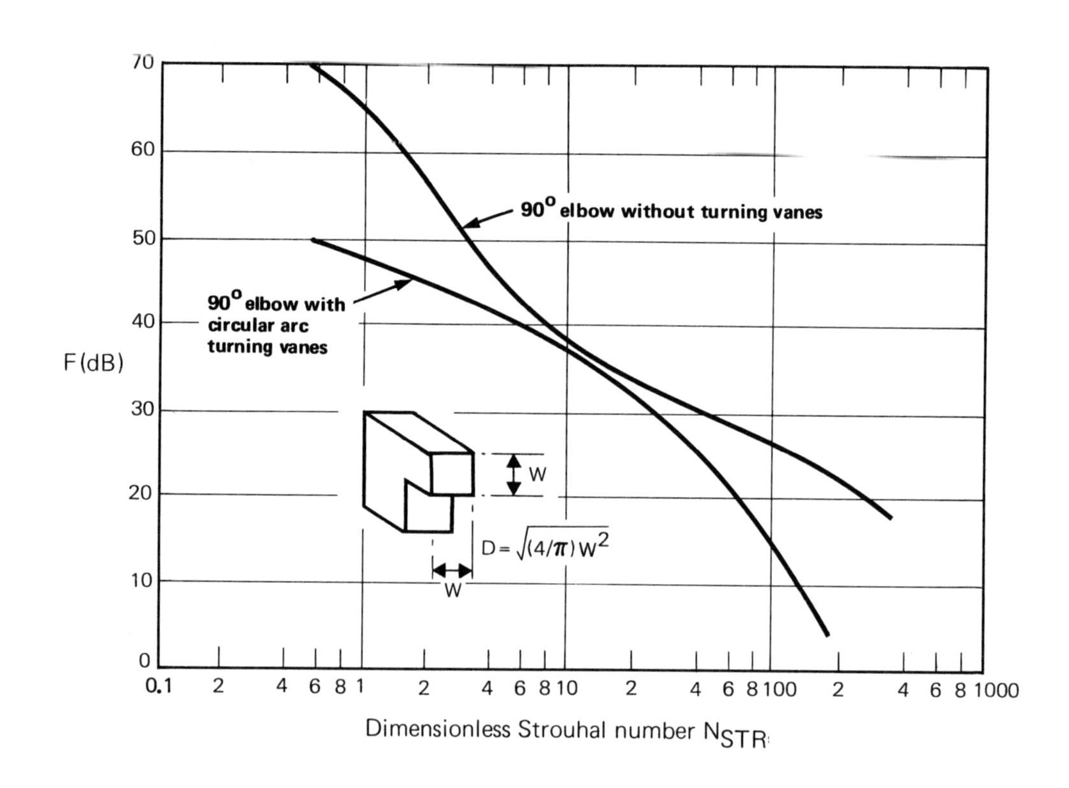

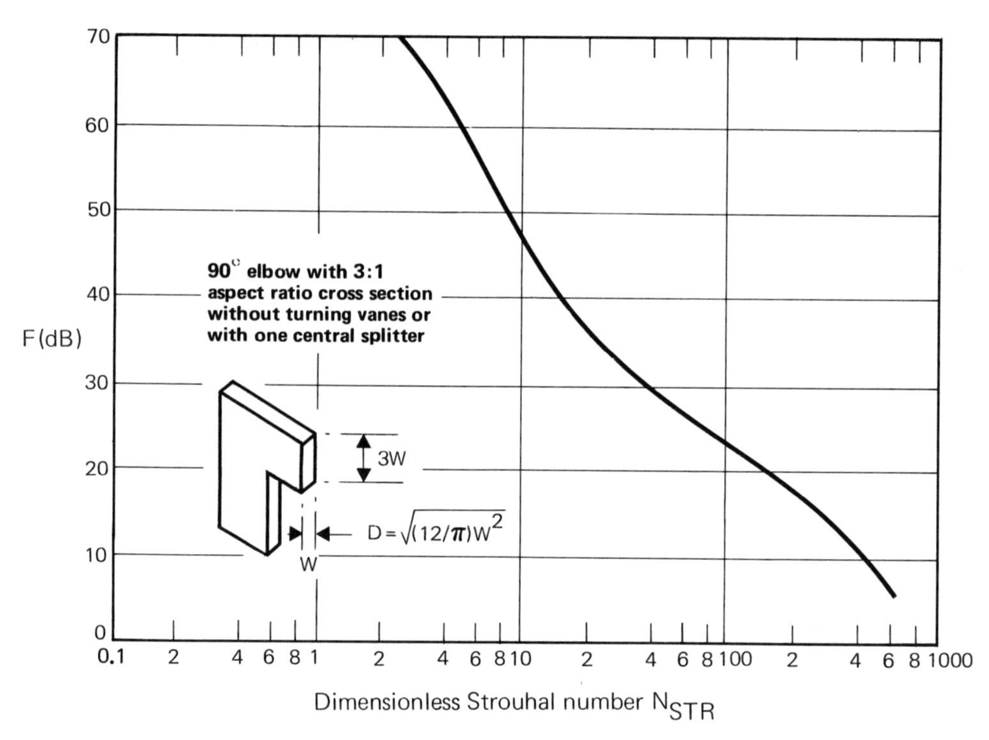

Based on data presented in ASHRAE Handbook (1973 Systems Volume ch 35)

FIG 3.2(b) 'G' factors for calculating velocity generated noise of duct elbows and grilles

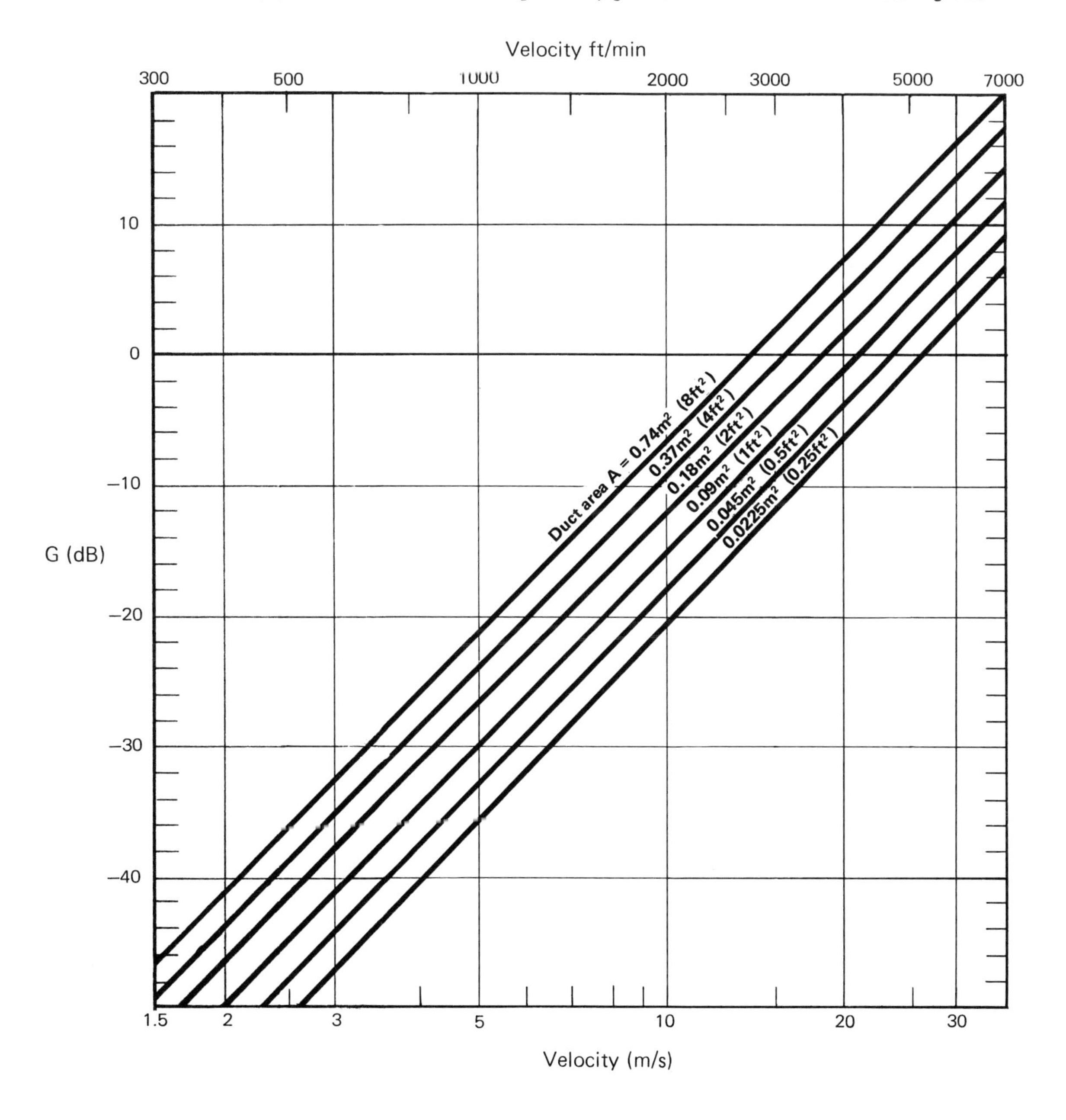

FIG 3.2(c) 'H' factors for calculating velocity generated noise of all duct fittings

Octave band (Hz)	63	125	250	500	1000	2000	4000	8000
H	16	19	22	25	28	31	34	37

Velocity generated PWL = F + G + H dB re 10^{-12} watts

Based on data presented in ASHRAE Handbook (1973 Systems Volume ch 35)

FIG 3.3(a) 'F' factors for calculating velocity generated noise in branch take-off ducts

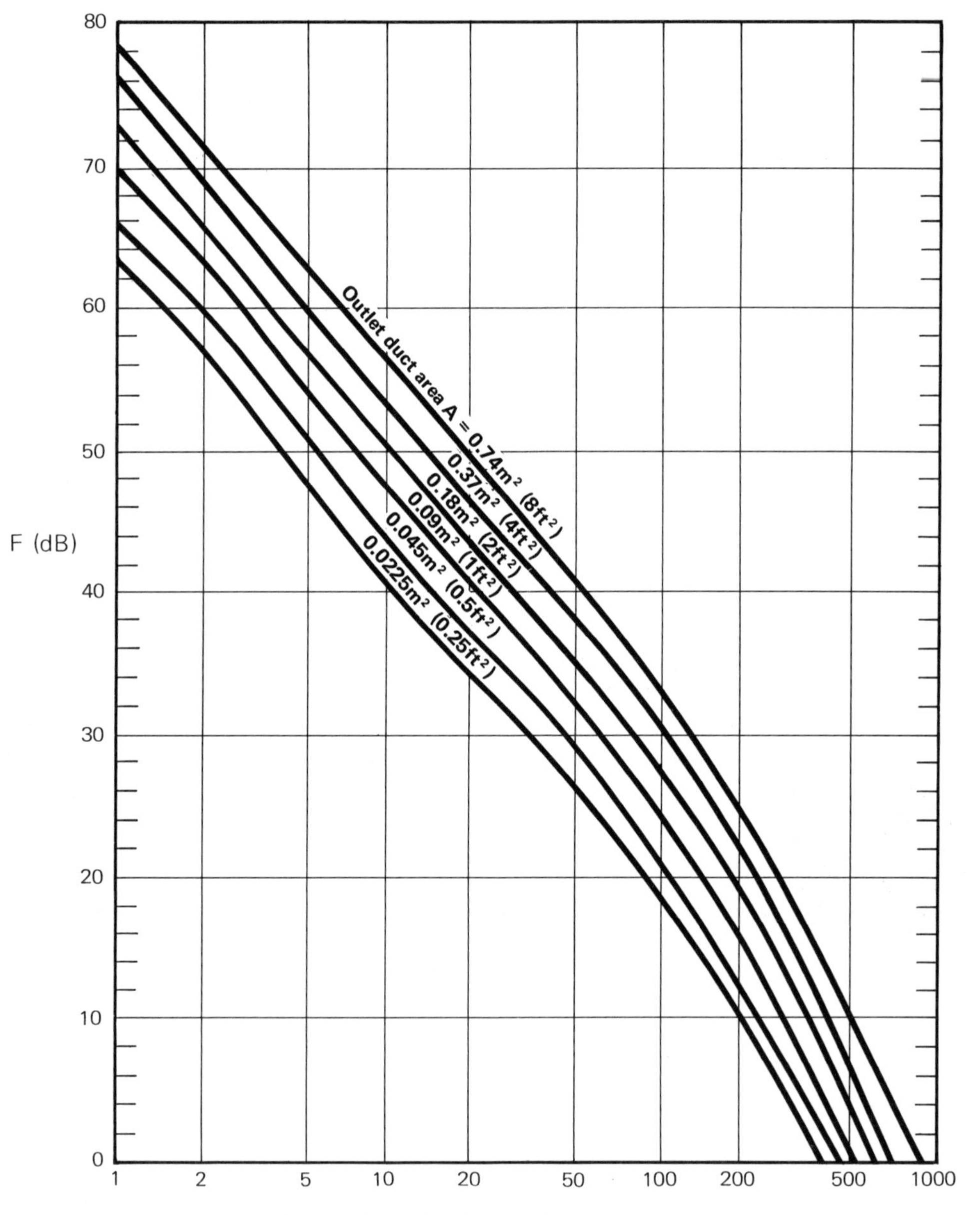

Velocity generated PWL = F + G + H dB re 10^{-12} Watts
G factor from Fig 3.3(b) opposite
H factor from Fig 3.2(c) on page

Based on data presented in ASHRAE Handbook (1973 Systems Volume ch 35)

FIG 3.3(b) (metric) 'G' factors for calculating velocity generated noise in branch take-off ducts

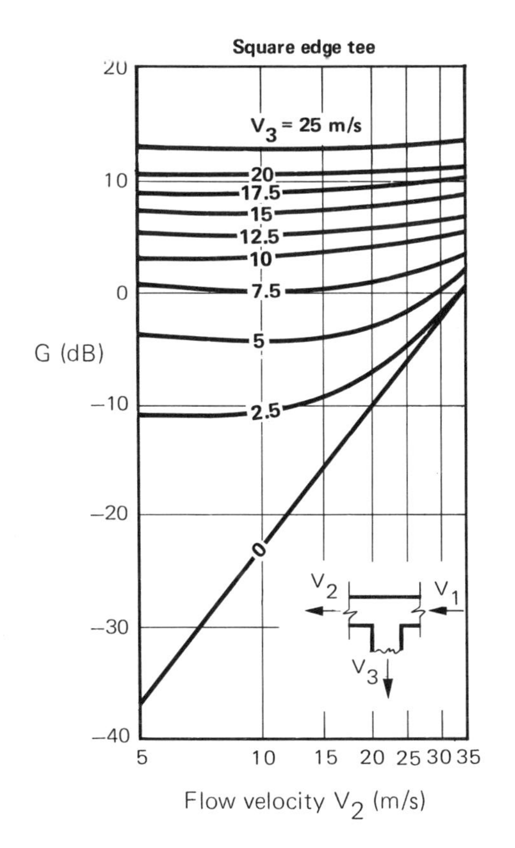

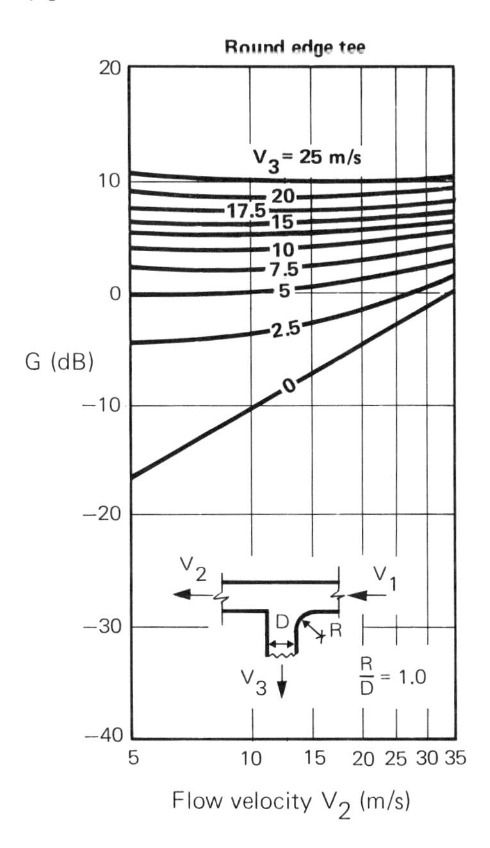

FIG 3.3(b) (Imperial) 'G' factors for calculating velocity generated noise in branch take-off ducts

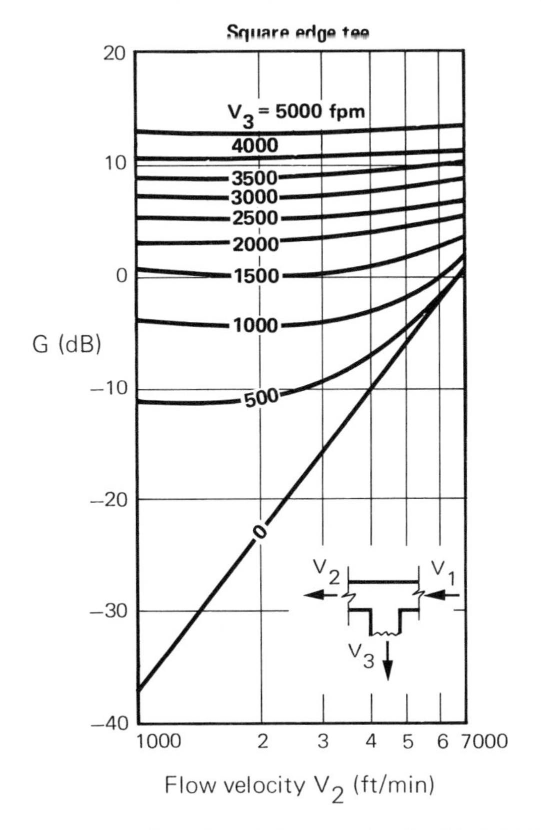

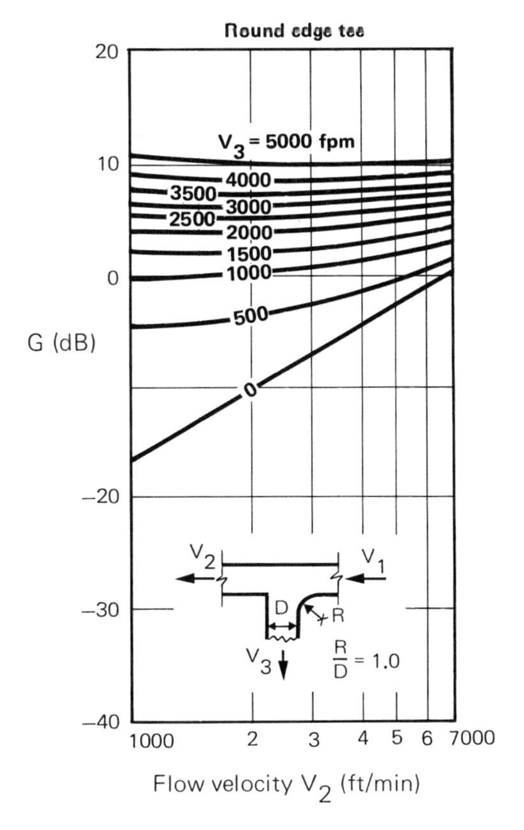

Based on data presented in ASHRAE Handbook (1973 Systems Volume ch 35)

FIG 3.4 'F' factors for calculating velocity generated noise of grilles

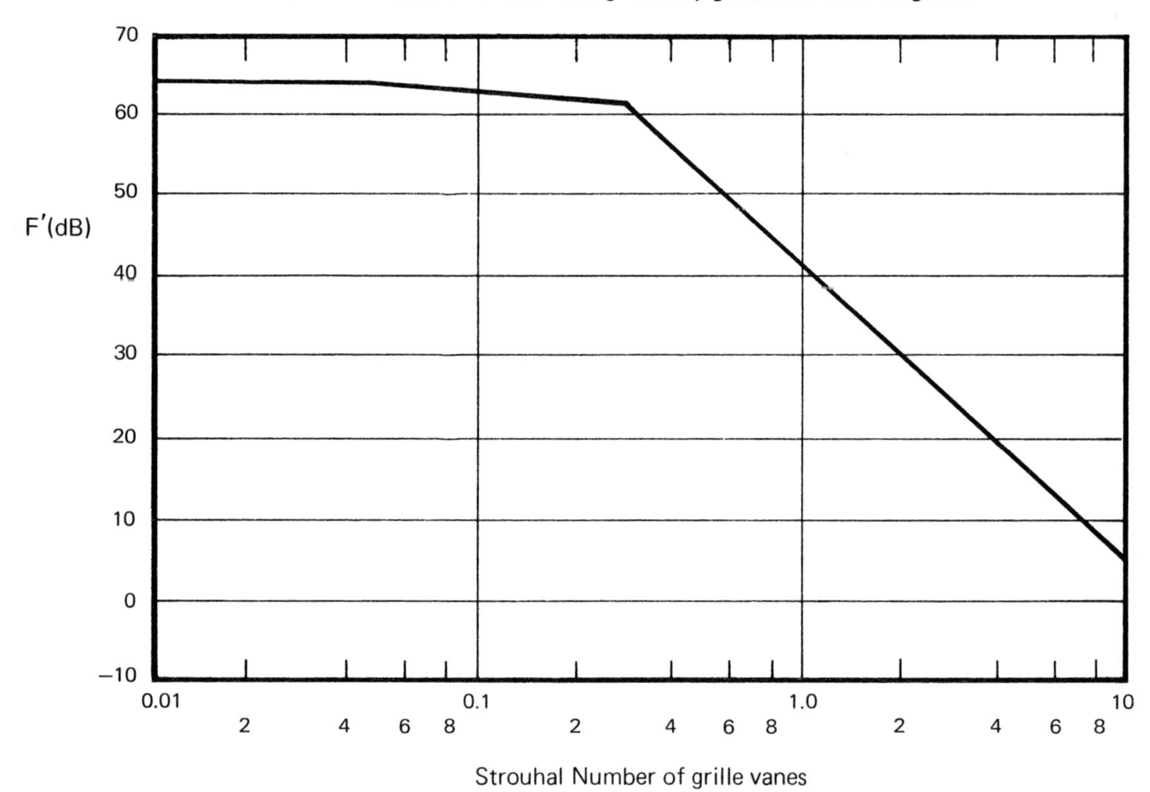

Strouhal Number of grille vanes

F = F' − K. Where K is determined by the thickness of the grille vanes.

d	in	.04	.08	.16	.32
	mm	1	2	4	8
K.		30	27	24	21

The Strouhal Number of the grille vanes is $\frac{f.d}{V_e}$ where:

V_e = Effective velocity $= \frac{cQ}{A_f}$

A_f = Free area of grille

Q = Airflow in duct

c = Constant = 1, for double deflect grilles
or = 2, for single deflect grilles

f = Frequency of octave band

d = Thickness of grille vanes

} in compatible units

Velocity generated PWL = F + G + H dB re 10^{-12} Watts

G factor from Fig 3.2(b)
H factor from Fig 3.2(c)

Based on HVRA Laboratory Report No. 75, 1973

FIG 3.5 Velocity generated noise in straight rectangular duct 600 mm x 600 mm (24 in x 24 in)

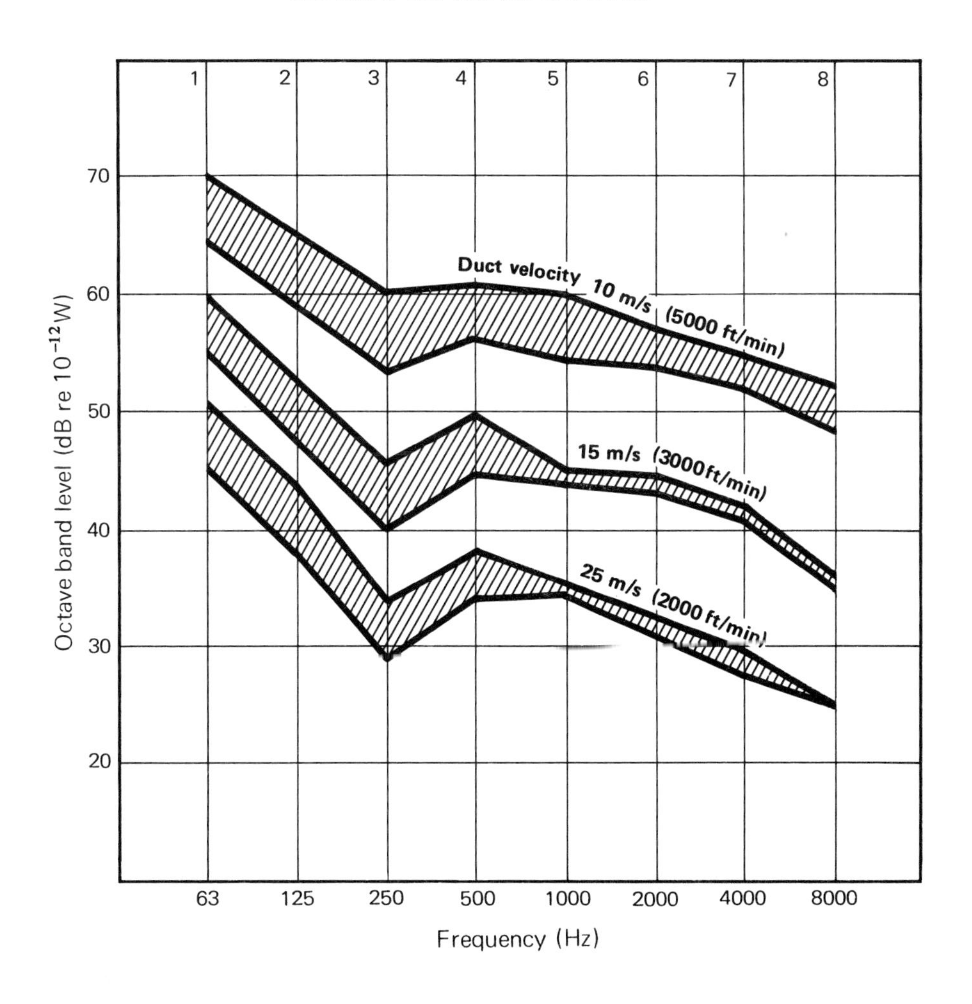

Based on data presented in ASHRAE Handbook (1973 Systems Volume ch 35)

FIG 3.6(a) (metric) Velocity generated noise of a typical damper—set at 0° (open) in a 600 mm x 600 mm duct

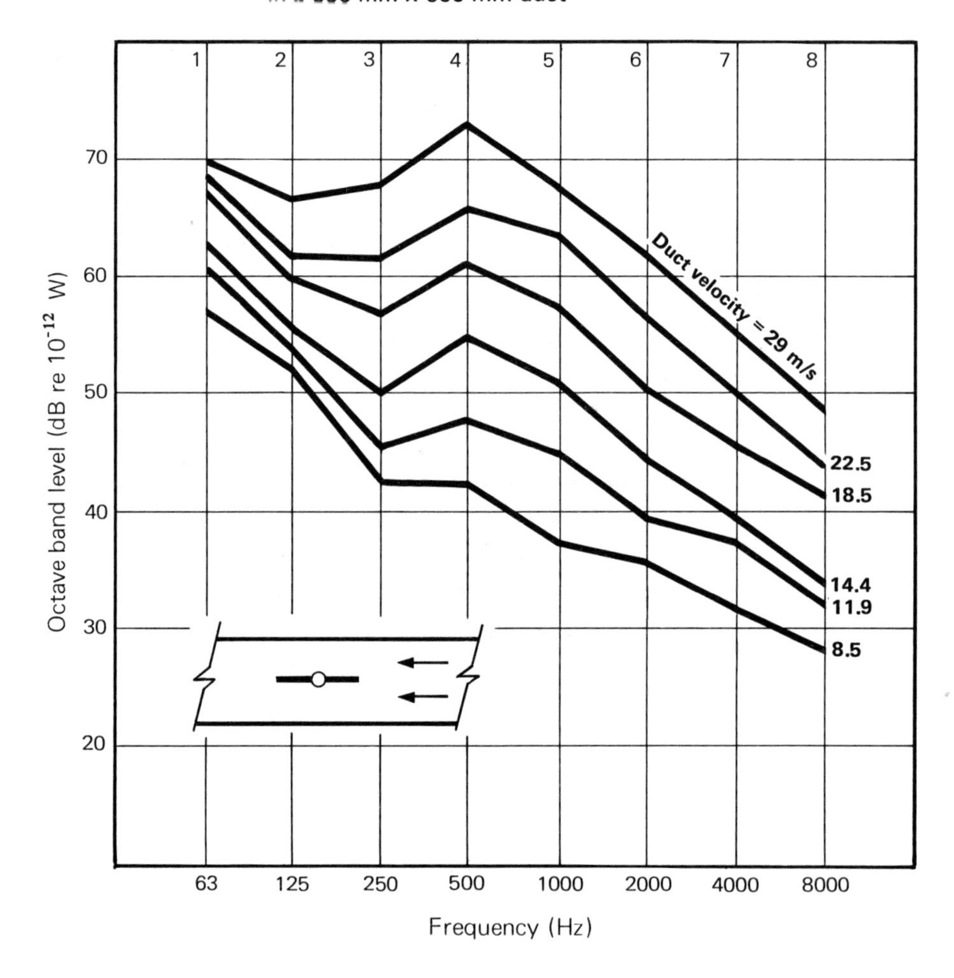

Based on data presented in ASHRAE Handbook (1973 Systems Volume ch 35)

FIG 3.6(a) (Imperial) Velocity generated noise of a typical damper—set at 0° (open) in a 24 in x 24 in duct

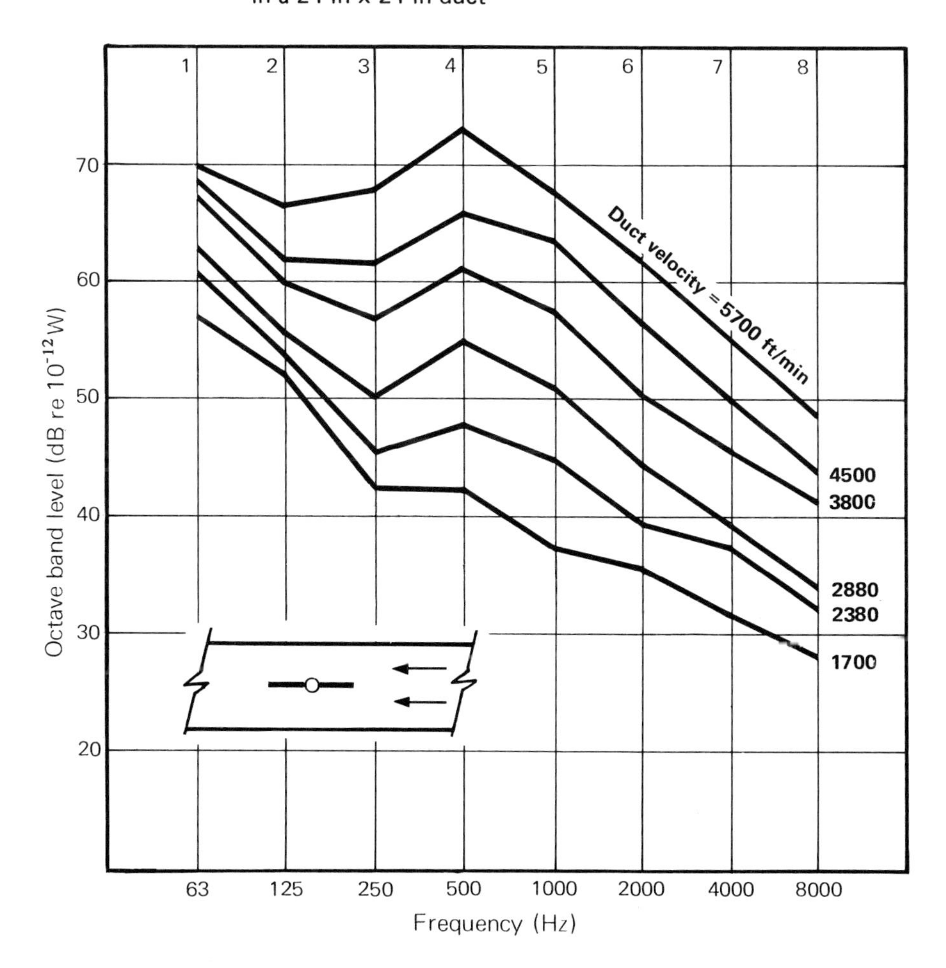

Based on data presented in ASHRAE Handbook (1973 Systems Volume ch 35)

FIG 3.6(b) (metric) Velocity generated noise of a typical damper—set at 15° in a 600 mm x 600 mm duct

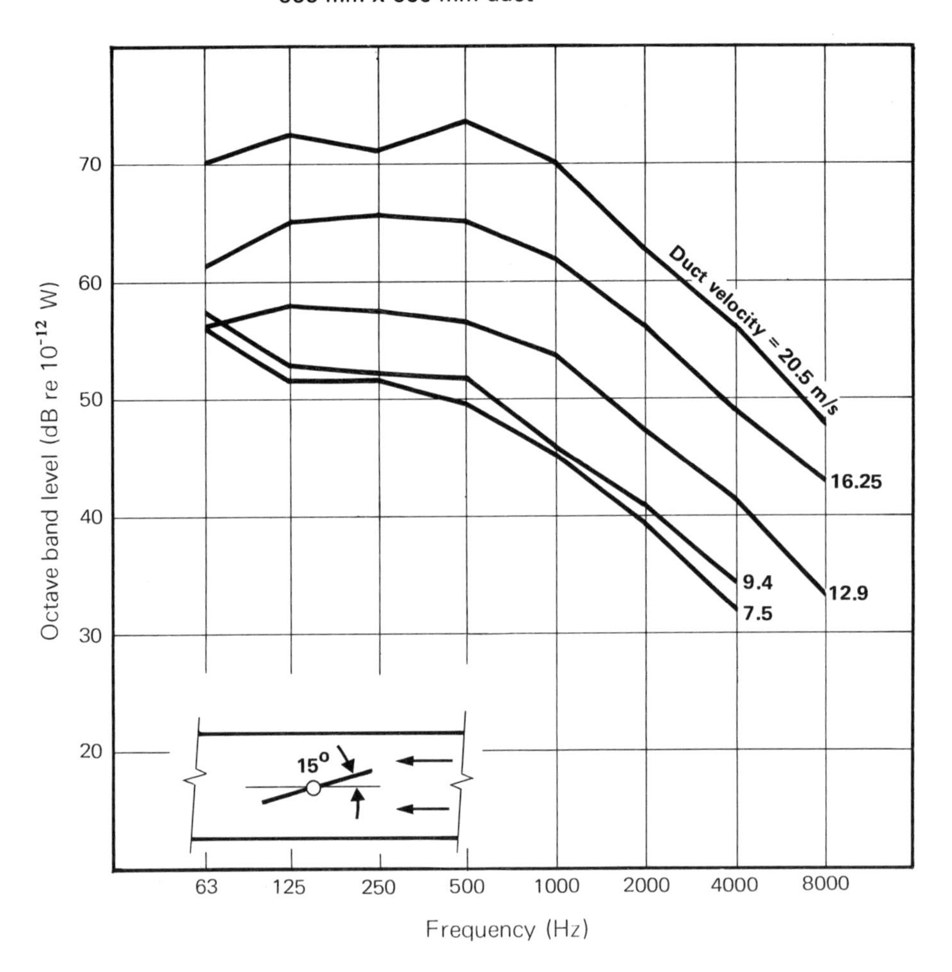

Based on data presented in ASHRAE Handbook (1973 Systems Volume ch 35)

FIG 3.6(b) (Imperial) Velocity generated noise of a typical damper—set at 15° in a 24 in x 24 in duct

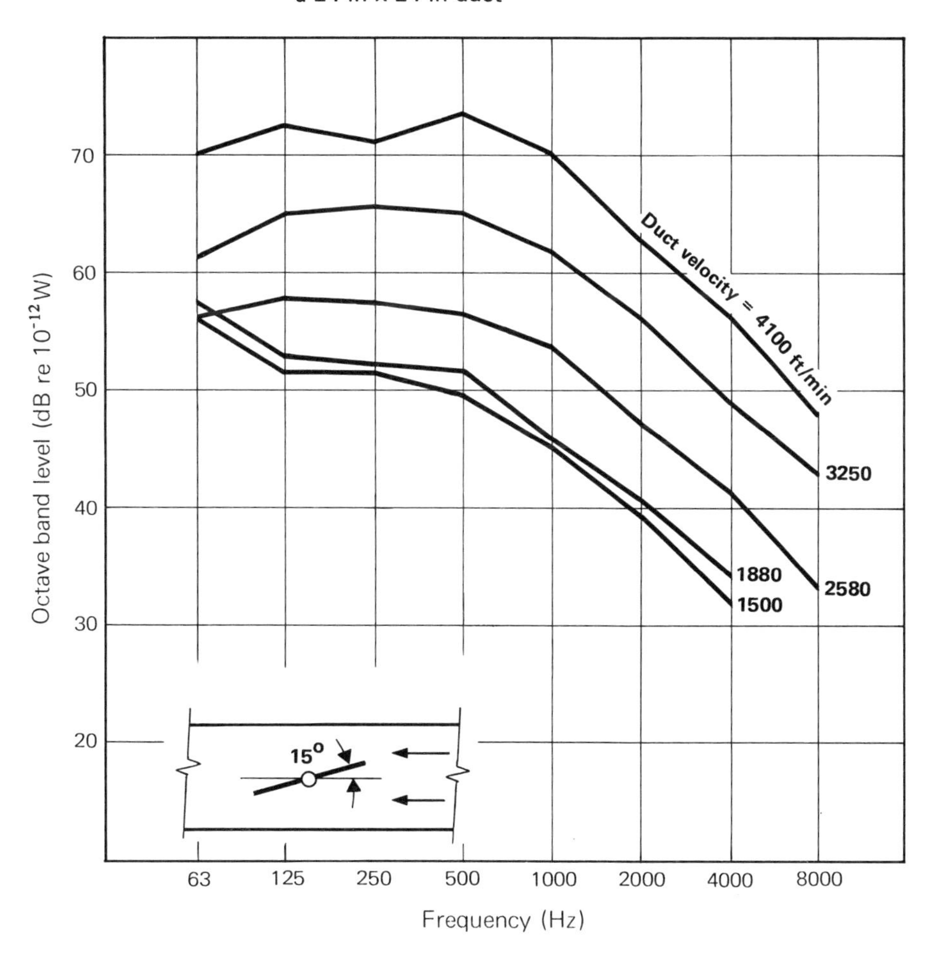

Based on data presented in ASHRAE Handbook (1973 Systems Volume ch 35)

FIG 3.6(c) (metric) Velocity generated noise of a typical damper—set at 45° in a 600 mm x 600 mm duct

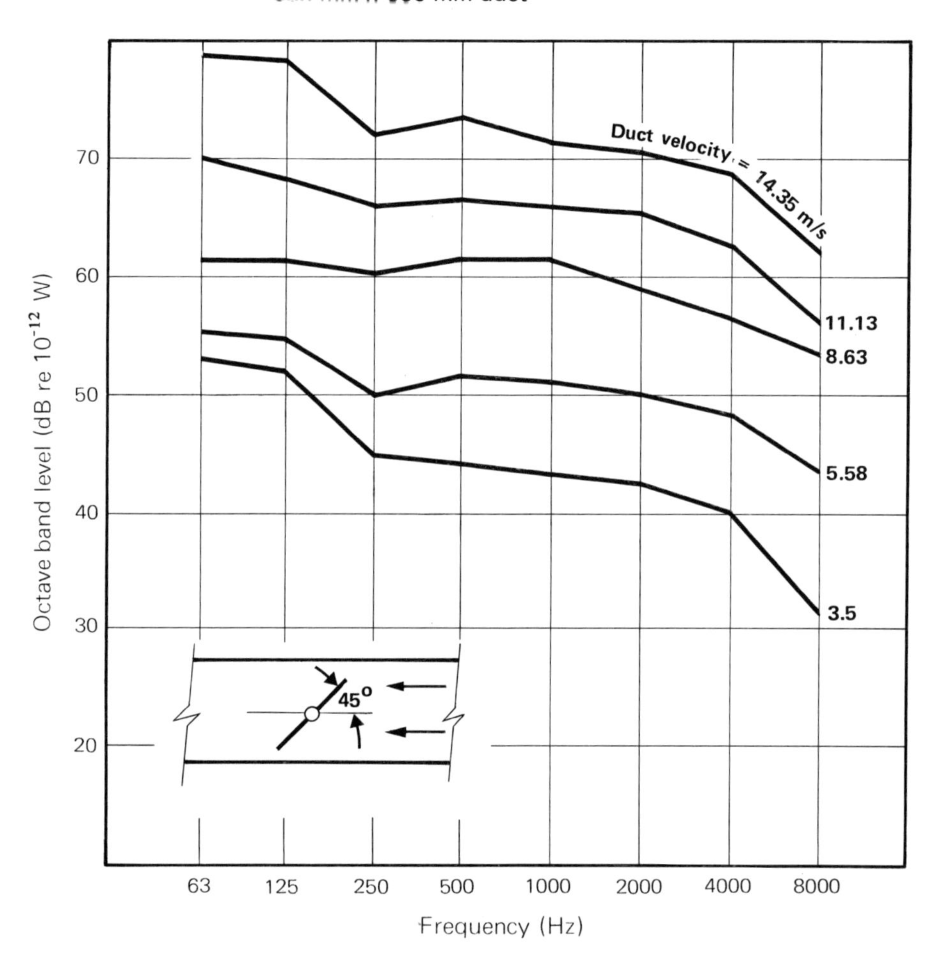

Based on data presented in ASHRAE Handbook (1973 Systems Volume ch 35)

FIG 3.6(c) (Imperial) Velocity generated noise of a typical damper—set at 45° in a 24 in x 24 in duct

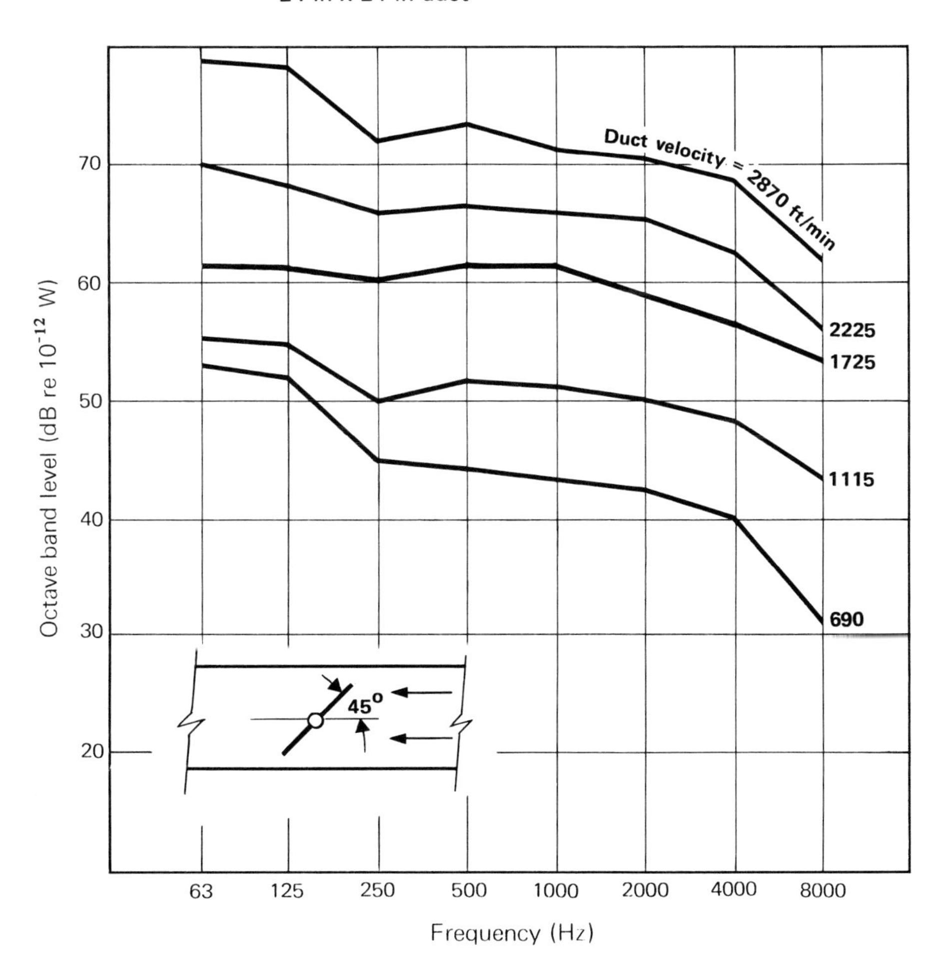

Based on data presented in ASHRAE Handbook (1973 Systems Volume ch 35)

FIG 3.7(a) (metric) Velocity generated noise of duct transition, 9:1

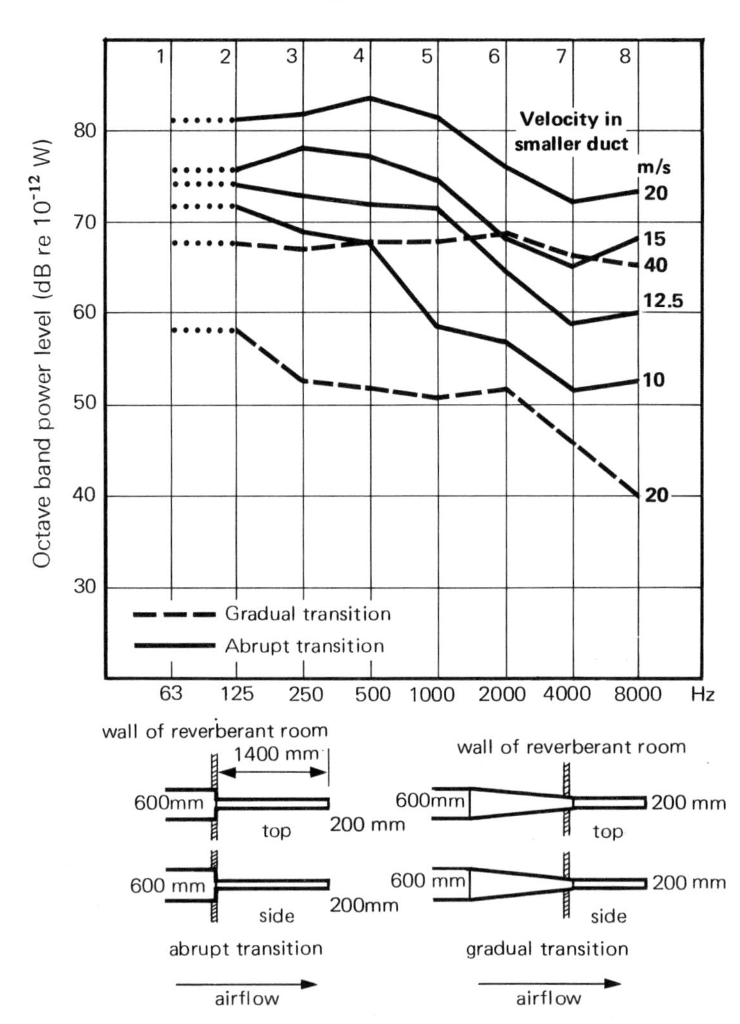

No data is available for 63 Hz band.
PWL values at 63 Hz are likely to be within ±5 dB of 125 Hz.
In the absence of exact data, it is recommended that 125 Hz values are also used for 63 Hz.

Based on data presented in ASHRAE Handbook (1973 Systems Volume ch 35)

FIG 3.7(a) (Imperial) Velocity generated noise of duct transition, 9:1

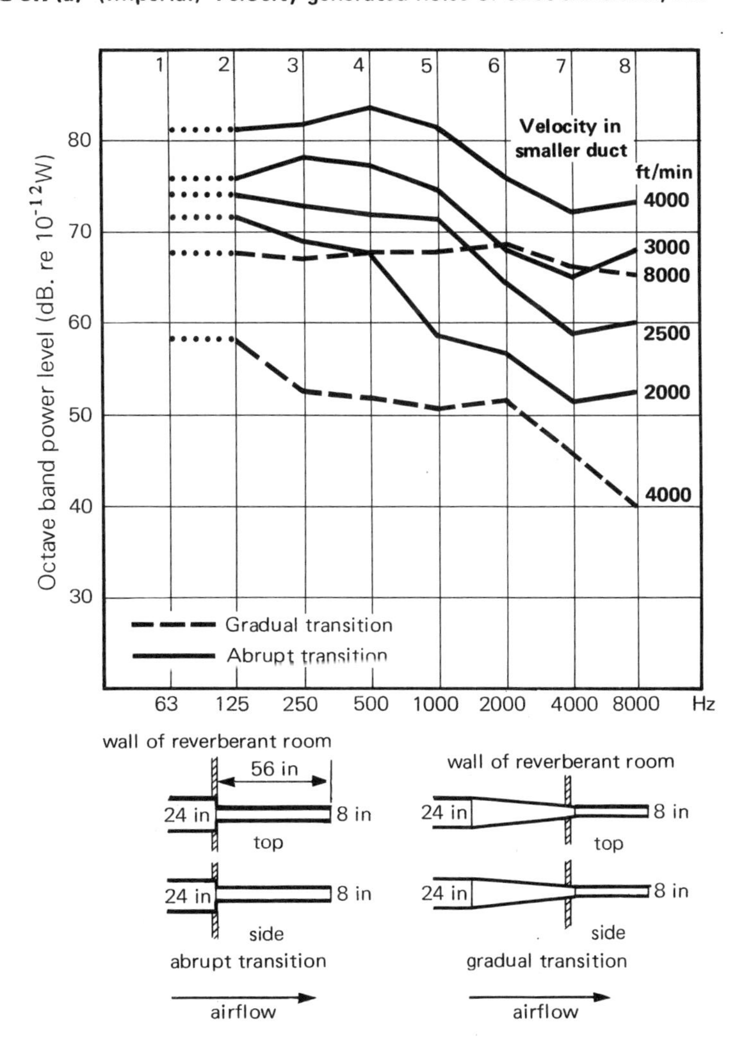

No data is available for 63 Hz band.
PWL values at 63 Hz are likely to be within ±5 dB of 125 Hz.
In the absence of exact data, it is recommended that 125 Hz values are also used for 63 Hz.

Based on data presented in ASHRAE Handbook (1973 Systems Volume ch 35)

FIG 3.7(b) (metric) Velocity generated noise of duct transition, 3:1

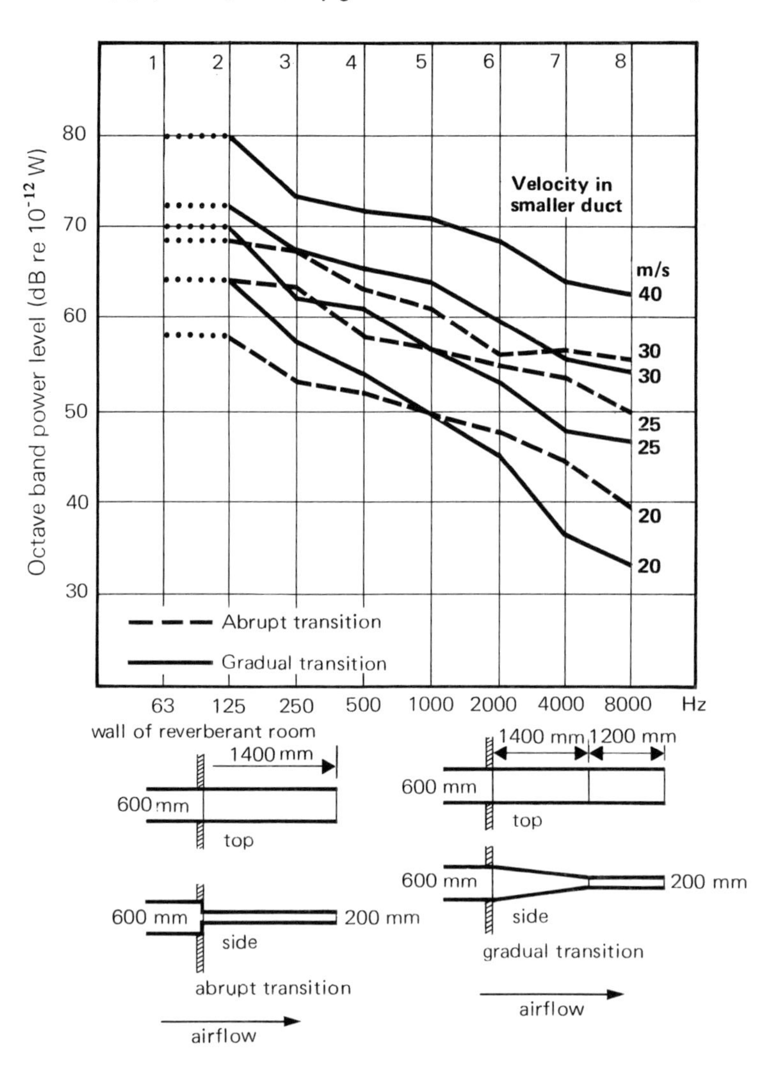

No data is available for 63 Hz band.
PWL values at 63 Hz are likely to be within ±5 dB of 125 Hz.
In the absence of exact data, it is recommended that 125 Hz values are also used for 63 Hz.

Based on data presented in ASHRAE Handbook (1973 Systems Volume ch 35)

FIG 3.7(b) (Imperial) Velocity generated noise of duct transition, 3:1

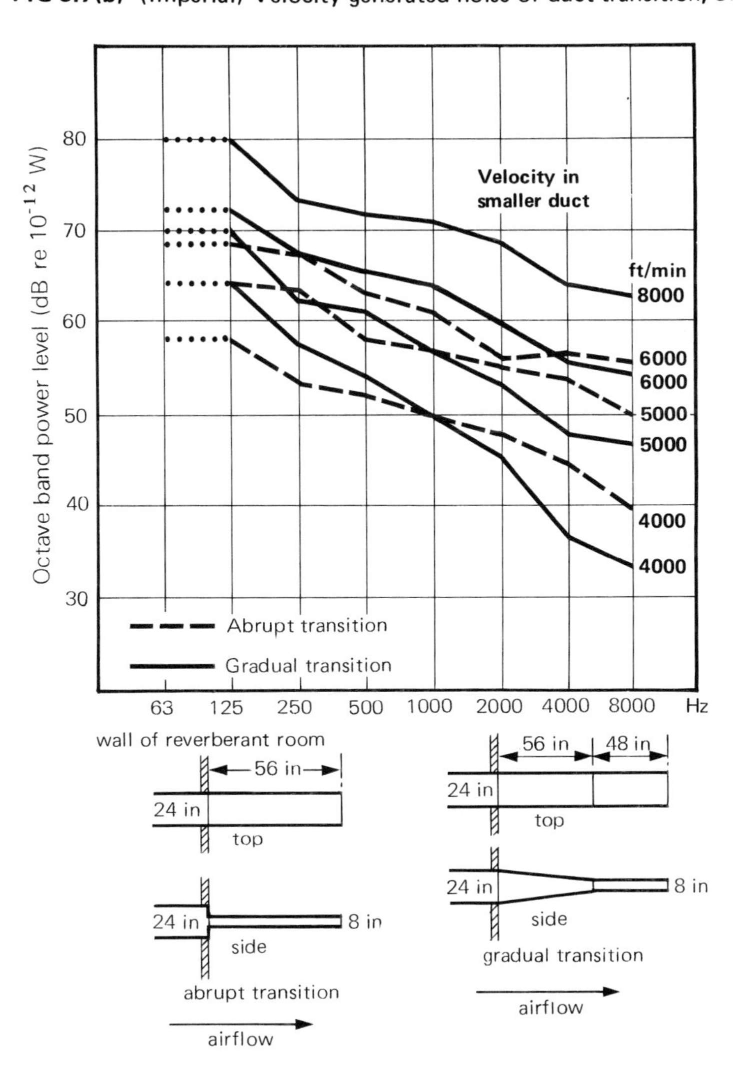

No data is available for 63 Hz band.
PWL values at 63 Hz are likely to be within ±5 dB of 125 Hz.
In the absence of exact data, it is recommended that 125 Hz values are also used for 63 Hz.

Based on data presented in ASHRAE Handbook (1973 Systems Volume ch 35)

FIG 3.7(c) (metric) Velocity generated noise of duct transition, 4:3

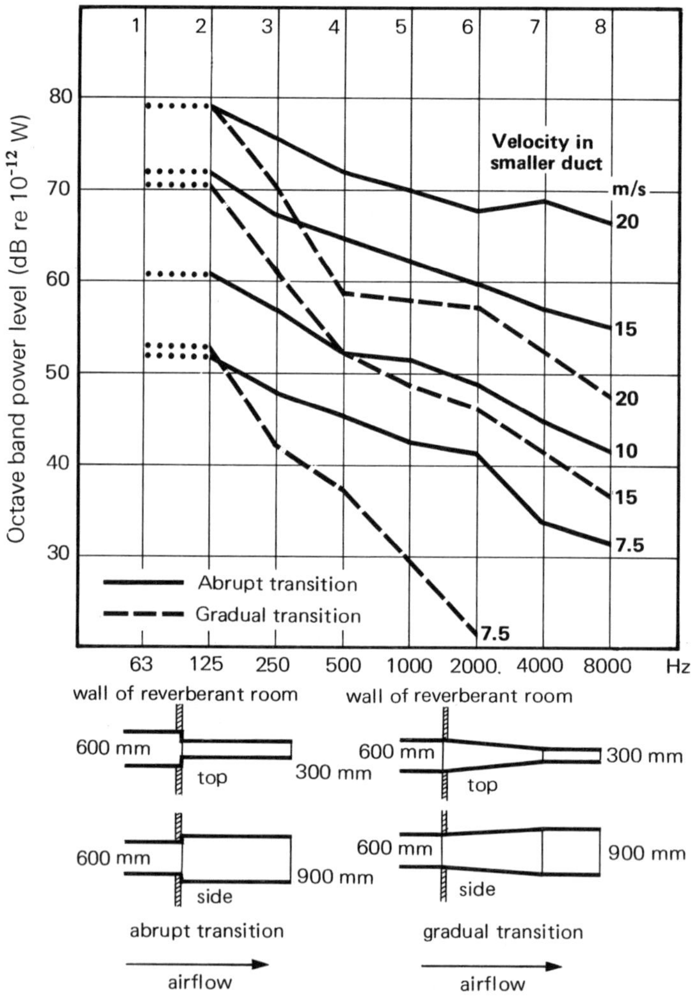

No data is available for 63 Hz band.
PWL values at 63 Hz are likely to be within ±5 dB of 125 Hz.
In the absence of exact data, it is recommended that 125 Hz values are also used for 63 Hz.

Based on data presented in ASHRAE Handbook (1973 Systems Volume ch 35)

FIG 3.7(c) (Imperial) Velocity generated noise of duct transition, 4:3

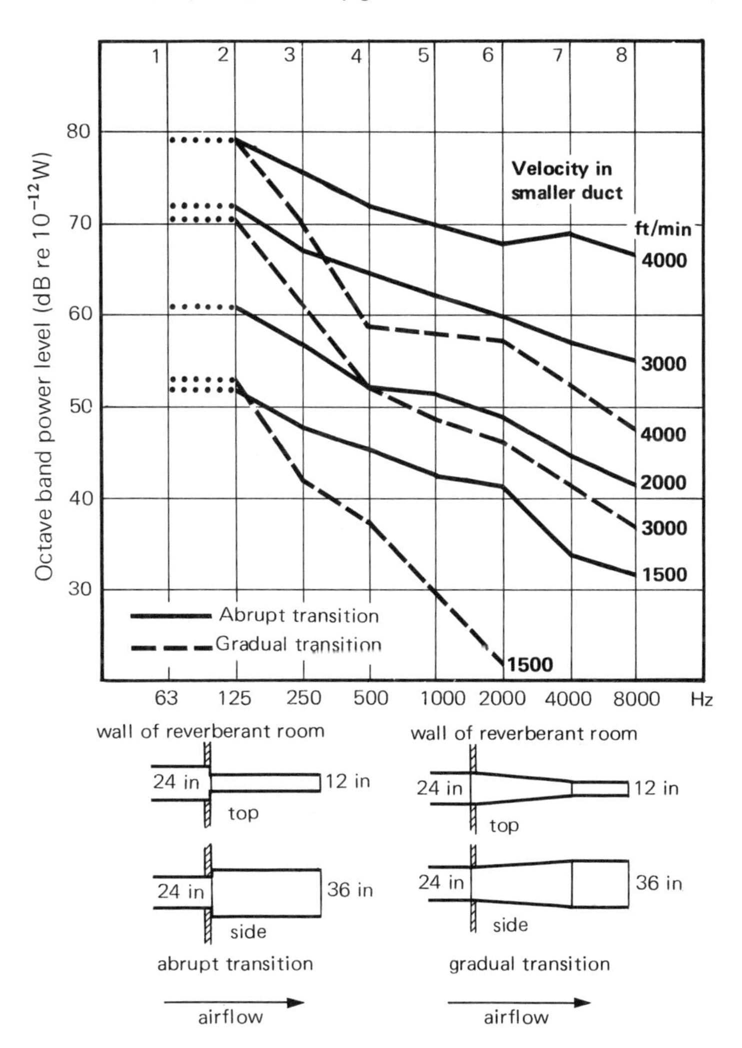

No data is available for 63 Hz band.
PWL values at 63 Hz are likely to be within ±5 dB of 125 Hz.
In the absence of exact data, it is recommended that 125 Hz values are also used for 63 Hz.

Based on data presented in ASHRAE Handbook (1973 Systems Volume ch 35)

FIG 3.7(d) (metric) Velocity generated noise of duct transition, 1.5:1

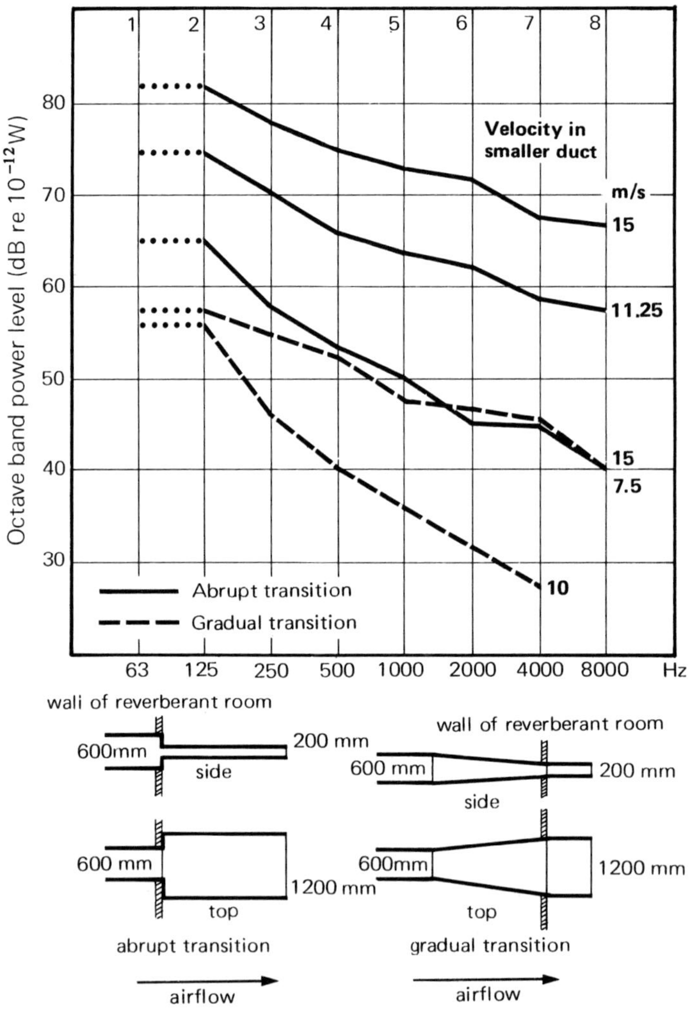

No data is available for 63 Hz band.
PWL values at 63 Hz are likely to be within ±5 dB of 125 Hz.
In the absence of exact data, it is recommended that 125 Hz values are also used for 63 Hz.

Based on data presented in ASHRAE Handbook (1973 Systems Volume ch 35)

FIG 3.7(d) (Imperial) Velocity generated noise of duct transition, 1.5:1

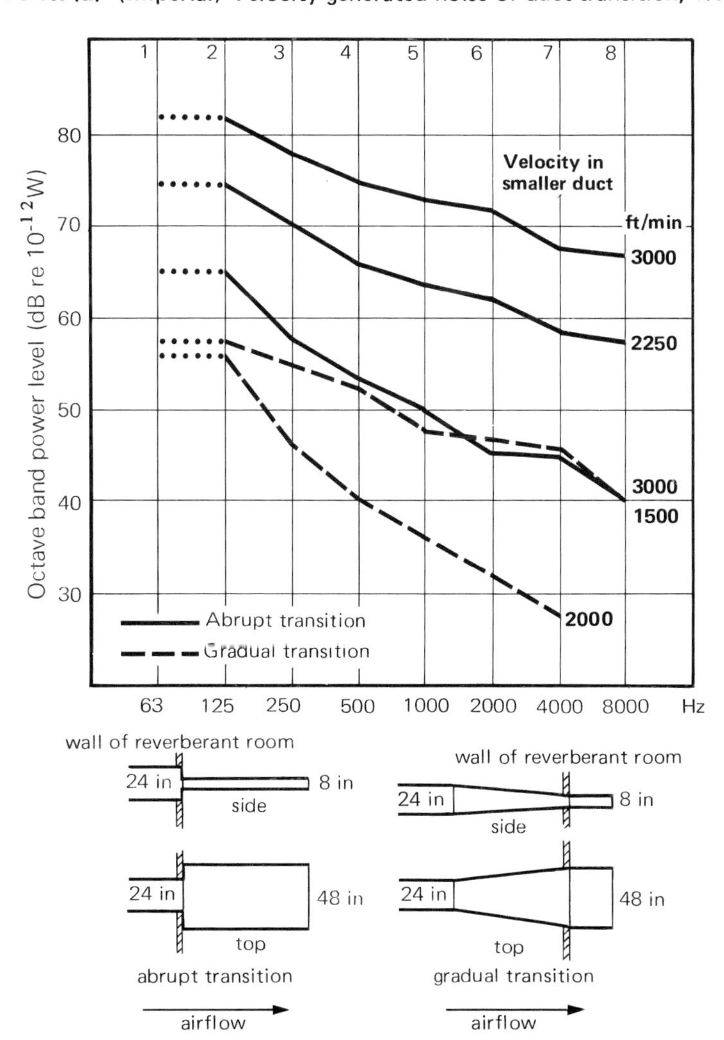

No data is available for 63 Hz band.
PWL values at 63 Hz are likely to be within ±5 dB of 125 Hz.
In the absence of exact data, it is recommended that 125 Hz values are also used for 63 Hz.

Based on data presented in ASHRAE Handbook (1973 Systems Volume ch 35)

FIG 3.8(a) Velocity generated noise of silencers

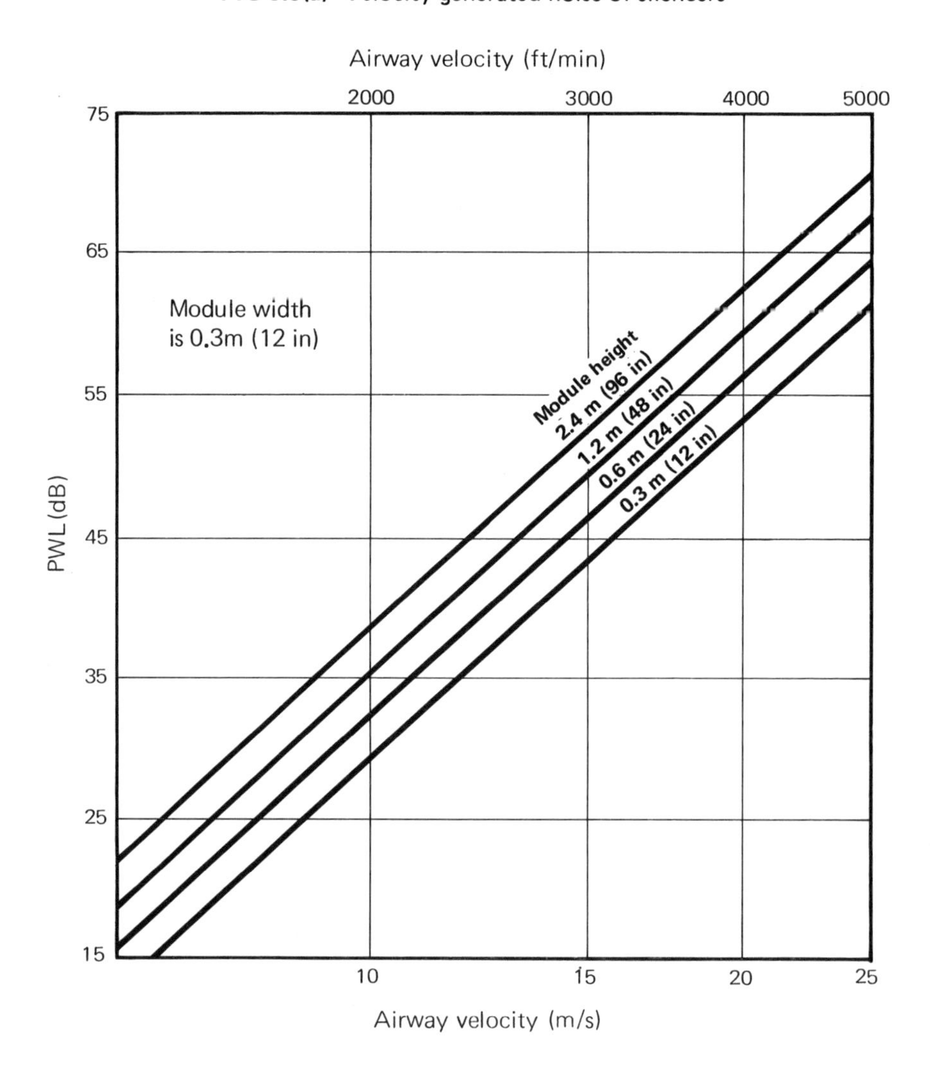

FIG 3.8(b) Spectrum correction

Octave band (Hz)	63	125	250	500	1000	2000	4000	8000
Correction (dB)	0	0	0	−4	−13	−13	−19	−22

FIG 3.8(c) Correction for number of modules

Number of modules	2	3	4	5	6	8	10
Correction (dB)	+3	+5	+6	+7	+8	+9	+10

Courtesy of Sound Attenuators Ltd.

CHAPTER FOUR

Duct breakout noise

4.1 Introduction.

Terminal outlets and grilles are not the only paths for the noise to enter the conditioned space. A duct carrying large volumes of air at high velocities can be subject to drumming, and radiate noise into the conditioned space. This is generally known as 'duct breakout noise'. Even in low velocity systems it is possible for the noise generated by the fan to break out of the ductwork and cause problems in areas over which it passes, regardless of whether those areas are served by the duct or not. A common situation arises when the fan noise breaks out of the duct into an area such as the plant room and gets back into the duct, just past the silencer, thereby short circuiting it. This is often termed as 'flanking', though flanking, strictly speaking, is not breakout noise but the noise travelling along the duct walls, bypassing the silencers. This problem is normally encountered when silencers in excess of 50 dB are used. Figure 4.1 illustrates diagrammatically two of the most common situations of duct breakout noise in low velocity systems. The problem is most severe in the first 3 or 6 m (10 or 20 ft) of duct where the induct power level is high, particularly if the duct passes over a critical room. A heavy density impervious false ceiling may help to reduce the breakout noise reaching the critical room.

4.2 Dealing with duct breakout noise.

In general, whenever the power level induct is high, or secondary silencers just upstream of the outlet are used, e.g. Fig. 4.1(a), or lengths of duct with large surface areas close to the fan pass over critical unconditioned areas, e.g. Fig. 4.1(b), the possibility of additional silencing must be looked at in order to reduce the power level in that duct. Breakout from a duct made of a good acoustic insulating material, such as concrete or brickwork, clearly will be appreciably less than that from a poor insulator such as light gauge steel. Lagging of ducts, therefore, appears to be a remedy, but in practice it works out considerably more expensive than reducing the induct power level by providing additional silencing.

The exact estimation of duct breakout noise is quite complex, and not yet fully understood. Research is being carried out to investigate it in greater detail. The best method available at present is the Allen formula, which predicts the duct breakout power level for a given power level in a duct. The breakout sound power results in a certain sound pressure level in the affected room, depending upon its size and absorption characteristics. In the calculation of the resultant sound pressure level in the room, the calculation procedure is identical to that used for duct-borne noise.

FIG 4.1 Duct breakout noise

(a) breakout in the presence of secondary silencers

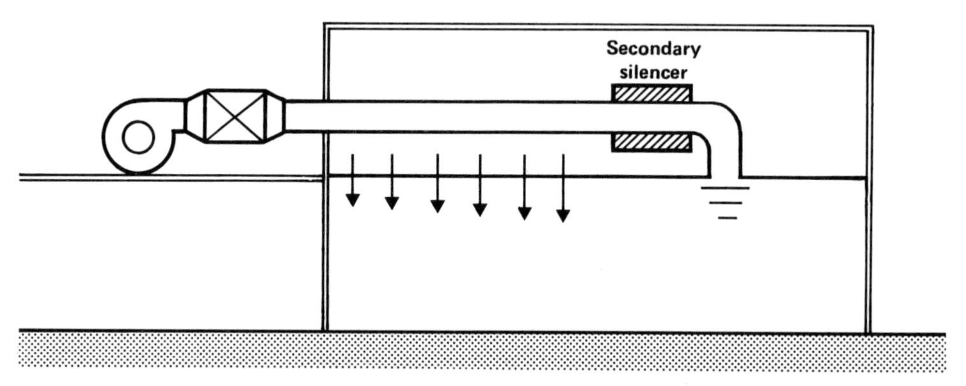

(b) breakout into an unconditioned space

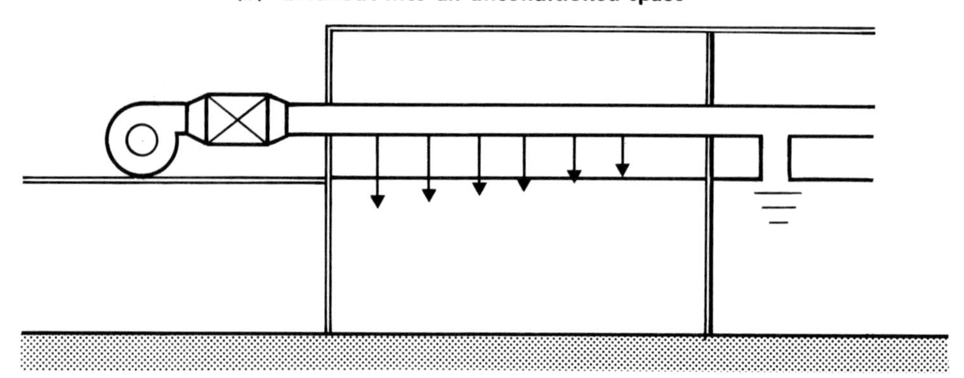

4.3 Allen formula.

Fan noise breakout is calculated from the sound power density being transmitted along the duct, by

$$PWL_B = PWL_D - R + 10 \log_{10} S_W/A$$

where:

PWL_B = the sound power level of noise breaking out of the duct wall, dB
PWL_D = the sound power level being transmitted along the duct, dB
R = the sound reduction index of the duct wall, dB
S_W/A = the ratio of the exposed surface area S_W to the cross-sectional area A (in consistent units).

For very large values of S_W/A, i.e. large areas of exposed ductwork, the value of $10 \log_{10} S_W/A$ can become greater than R, which implies that more sound power comes out of the duct than was inside originally. This is clearly impossible and it is therefore necessary to limit the value of PWL_B to at least 3 dB below PWL_D, i.e. to assume that half the sound power is transmitted along the duct and half escapes through the walls.

It is normal practice to consider only the first 9 m (30-ft) of a length of ductwork of any system, by which time the induct PWL would have dropped approximately 6 dB at low frequencies. Any length of ductwork beyond this would not contribute significantly to the estimate of the breakout noise problem.

EXAMPLE 1:

In the scheme shown in Fig. 2.1 (pp.13 and 15) estimate (i) the duct breakout noise where it may be critical; and (ii) calculate the additional silencing necessary. The director's office has a plasterboard ceiling, giving a transmission loss of 5, 11, 17, 23, 27, 30, 33, 32 dB in each octave band respectively.

SOLUTION

In this scheme, the branch AJK pases over the most critical area (NC 35) and is carrying almost 60 per cent of the total airflow. Breakout noise is therefore likely to be a problem in this section.

The analysis chart in Fig. 4.2 shows, step by step, the calculation required to reach a figure for the SPL due to breakout and for the silencing required. The example in section 4.4 evaluates suitable silencing options to deal with this problem.

FIG. 4.2 Breakout noise analysis chart

	63	125	250	500	1000	2000	4000	8000 Hz
a Induct PWL at point A (before silencing)	82	84	84	82	81	80	76	70
b Using Allen formula $PWL_B = PWL_D - R + 10 \log_{10} S_W/A$ dB Breakout PWL * At frequencies 63, 125, 250 Hz, more sound PWL or nearly as much sound PWL as induct PWL appears to break out. This is impossible (see section 4.3) and we restrict breakout to ½ of induct PWL (i.e. less 3 dB)	92*	89*	82*	74	68	62	52	47
c Corrected breakout PWL	79	81	81	74	68	62	52	47
d Transmission loss of false ceiling	5	11	17	23	27	30	33	32
e Breakout PWL entering room (*c*–*d*)	74	70	64	51	41	32	19	15
Corrections for reverberant SPL								
f Room volume correction for 140 m³ (5000 ft³)	-8	-8	-8	-8	-8	-8	-8	-8
g Correction for 1½ s rev. time	+2	+2	+2	+2	+2	+2	+2	+2
h +3 dB correction for breakout from extract	+3	+3	+3	+3	+3	+3	+3	+3
i Total correction (*f* + *g* + *h*)	-3	-3	-3	-3	-3	-3	-3	-3
j Reverberant SPL (*e* + *i*)	71	67	61	48	38	29	16	12
Corrections for direct SPL								
k Correction for distance (assumed 1.5 m (5 ft))	-15	-15	-15	-15	-15	-15	-15	-15
l Correction for directivity (underside of duct) area 1·90 m² (2160 in²)	+6	+7	+8	+8	+9	+9	+9	+9
m +3 dB correction for breakout for extract duct	+3	+3	+3	+3	+3	+3	+3	+3
n Total correction (*k* + *l* + *m*)	-6	-5	-4	-4	-3	-3	-3	-3
o Direct SPL (*e* + *n*)	68	65	60	47	38	29	16	12
p Resultant SPL in room due to breakout (logarithmic sum (*j* + *o*))	73	69	64	51	41	32	19	15
q Criterion NC 35	60	52	45	40	36	34	33	32
Silencing required for breakout	13	17	19	11	5	0	0	0

4.4 Duct lagging.

Duct lagging as a means of reducing breakout is generally very costly, and often difficult to install.

However, ducts carrying supply air often require thermal insulation. Under such circumstances, advantage may be taken of the insulation, which can effect significant reduction of breakout in the low frequencies.

Given below are some typical duct treatments and their estimated transmission losses.

	63 Hz	125 Hz	250 Hz
(1) Duct covered with 50 mm (2 in) light density 32 kg/m^3 (2 lb/ft^3) mineral wool or fibreglass, with outer skin of 22 g aluminium, sheet steel or 6 mm (¼ in) Keene's cement.	5	10	20
(2) Duct covered with 50 mm (2 in) medium density 96 kg/m^3 (6 lb/ft^3) mineral wool or fibreglass with outer skin of 16 g aluminium, sheet steel or 12 mm (½ in) Keene's cement.	10	15	25
(3) Duct covered with 75 mm (3 in) heavy density 192 kg/m^3 (12 lbs/ft^3) mineral wool or fibreglass with outer skin of 16 g aluminium, sheet steel or 12 mm (½ in) Keene's cement.	15	20	30

In many cases, the breakout noise at low frequency exceeds the noise level transmitted via the grilles. When this happens, the silencer will be governed by the low frequency breakout requirements. Frequently, a silencer large enough to silence the breakout noise cannot be accommodated, and therefore a combination of silencer and duct lagging provides the only solution.

EXAMPLE 2:

For the grille borne and duct breakout noise analyses shown in Figs. 2.3 and 4.2, evaluate suitable silencing options.

SOLUTION

	63	125	250	500	1000	2000	4000	8000	Hz
Silencer requirements for grille borne noise (from Fig. 2.3)	0	1	16	7	0	4	12	10	(a)
Silencer requirements for breakout noise (from Fig. 4.2)	13	17	19	11	5	0	0	0	(b)
Options									
(i) From Appendix 3 it can be seen that to deal with both grille borne and breakout noise a 2.4 m (96 in) long silencer would be necessary. Even this, at 63 Hz, would theoretically be inadequate as the insertion loss is:	8	10	32	47	50	50	37	25	(c)
(ii) As an alternative a silencer may be selected for the grille borne noise only. For this a 1.2 m (48 in) long silencer would give sufficient attenuation:	5	9	16	27	32	30	21	15	(d)
The amount of breakout silencing would therefore be reduced to (*b* – *d*)	8	8	3	0	0	0	0	0	
This can be supplied by lagging (see section 4.4)									

4.5 Computer program for duct breakout.

A computer program is available which calculates the duct breakout PWL, for a given induct power level spectrum, and analyses its effect on a given room. It also calculates the amount of silencing necessary to reduce the induct PWL to meet a given NC requirement. For full details refer to the description of programs in Appendix 5.

CHAPTER FIVE

Cross talk, location of silencers

5.1 Cross talk

Sound power can get into ducts just as easily as it can come out. If two rooms are therefore served by a common duct, the problem can arise of sound from one room passing into the ductwork and out into the adjoining room. The noise could simply be raised speech or noise of machinery, e.g. typewriters, etc. For the former it is often only necessary to eliminate the higher frequencies, so that the speech becomes unintelligible, thus maintaining privacy between two rooms. A suitable method of providing attenuation within the duct between two rooms is therefore necessary.

It is essential to note that the purpose of putting any such silencing in the duct can be defeated if the partition wall between two adjacent rooms has a lower transmission loss than the insertion loss of the silencer. Transmission losses of various materials are given in Appendix 1. The figures shown are those obtained under laboratory conditions, and may not be achieved in normal everyday installations.

For average installations, transmission losses are about 5dB lower than that indicated by the manufacturer of the material. Double ended electrical fittings in partition walls or electrical conduits with ends in two adjacent rooms can reduce the effectiveness of the partition considerably.

It is obvious that the lower the NC level in a room, the more chance there is of hearing speech or other noise from adjacent rooms via the connecting ductwork. Consequently a greater degree of cross talk silencing is required. The most common practice of providing cross talk attenuation is by putting purpose-made cross talk silencers in the interconnecting duct.

It is often possible to eliminate cross transmission of noise just by utilising the natural attenuation of the duct system, for example by extending the duct length. Figure 5.1 (a) and (b) shows a revised duct scheme which is just one way of achieving this.

Even a relatively short duct length having a few bends in the ductwork between the two outlet terminals can provide speech privacy, provided the ductwork is separated from the room (e.g. in a heavy suspended ceiling) so that noise does not enter the duct via its exposed surface.

FIG 5.1 Duct schemes

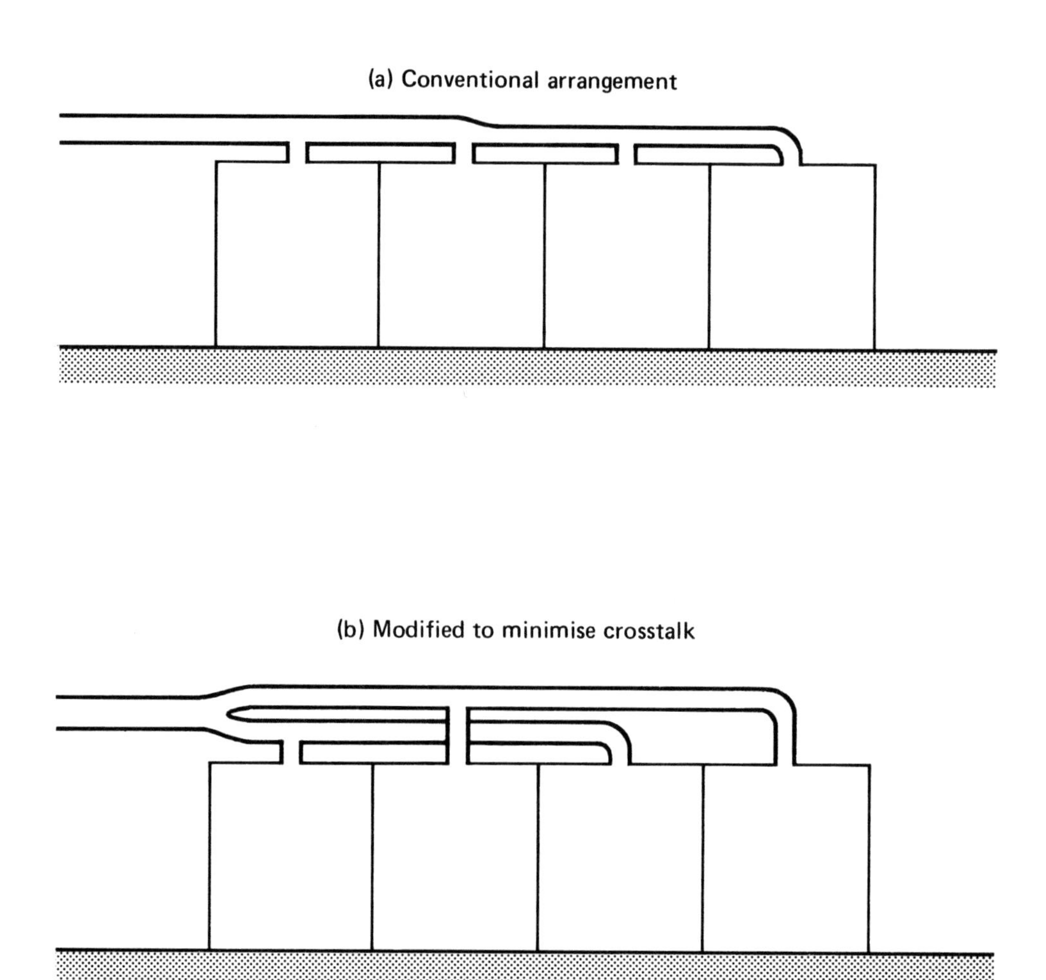

5.2 Selection of cross talk silencers.

The following method will enable a quick selection of suitable cross talk silencers which will provide adequate attenuation of raised speech and common office noise, e.g. typewriters, etc. for a required NC level down to NC 30 in the receiving room.

The fundamental silencer types are classified into six groups, A, B, C, D, E and F. The dimensions of each group are given in Table 5.1. Each is based on a module size 300 mm (12 in) high and having a 75 mm (3 in) free airway, module heights can be increased in 75 mm (3 in) steps up to 530 mm (21 in).

TABLE 5.1 Silencer dimensions

Group	*Attenuation dB*	*Length (m)*	*(ft)*	*Lining thickness (mm)*	*(in)*	*O/A width (mm)*	*(in)*
A	5	0.6	2	50	2	150	7
B	10	0.9	3	50	2	150	7
C	15	1.2	4	50	2	150	7
D	20	1.5	5	50	2	150	7
E	25	1.2	4	75	3	225	9
F	30	1.5	5	75	3	225	9

Table 5.2 gives the group of silencer that will be required to meet a given NC level in a receiving room, together with the maximum airflows permissible through each module. The airflows given are the maximum permissible through the silencer beyond which the velocity generated noise could become excessive.

TABLE 5.2 (metric) Silencer selection table

Required NC level in receiving room	*Type of silencer*	*Maximum allowable airflow in duct, m^3/h*			
		One 300 mm module	*Two 300 mm modules*	*Three 300 mm modules*	*Four 300 mm modules*
55	A	1 700	2 980	4 210	5 100
50	B	1 490	2 550	3 570	4 250
45	C	1 280	2 120	2 930	3 400
40	D	1 060	1 700	2 300	2 720
35	E	850	1 530	1 910	2 210
30	F	765	1 280	1 530	1 700

TABLE 5.2 (Imperial) Silencer selection table

Required NC level in receiving room	*Type of silencer*	*Maximum allowable airflow in duct, ft^3/min*			
		One 12 in module	*Two 12 in modules*	*Three 12 in modules*	*Four 12 in modules*
55	A	1 000	1 750	2 480	3 000
50	B	875	1 500	2 100	2 500
45	C	750	1 250	1 725	2 000
40	D	625	1 000	1 350	1 600
35	E	500	900	1 125	1 300
30	F	450	750	900	1 000

Air pressure drops must be checked from Fig. 5.2. If they are found excessive, increase the module height or the number of modules which will allow the desired air volume flow without exceeding a tolerable pressure drop limit.

FIG 5.2 (metric) Air pressure drop through cross talk silencers

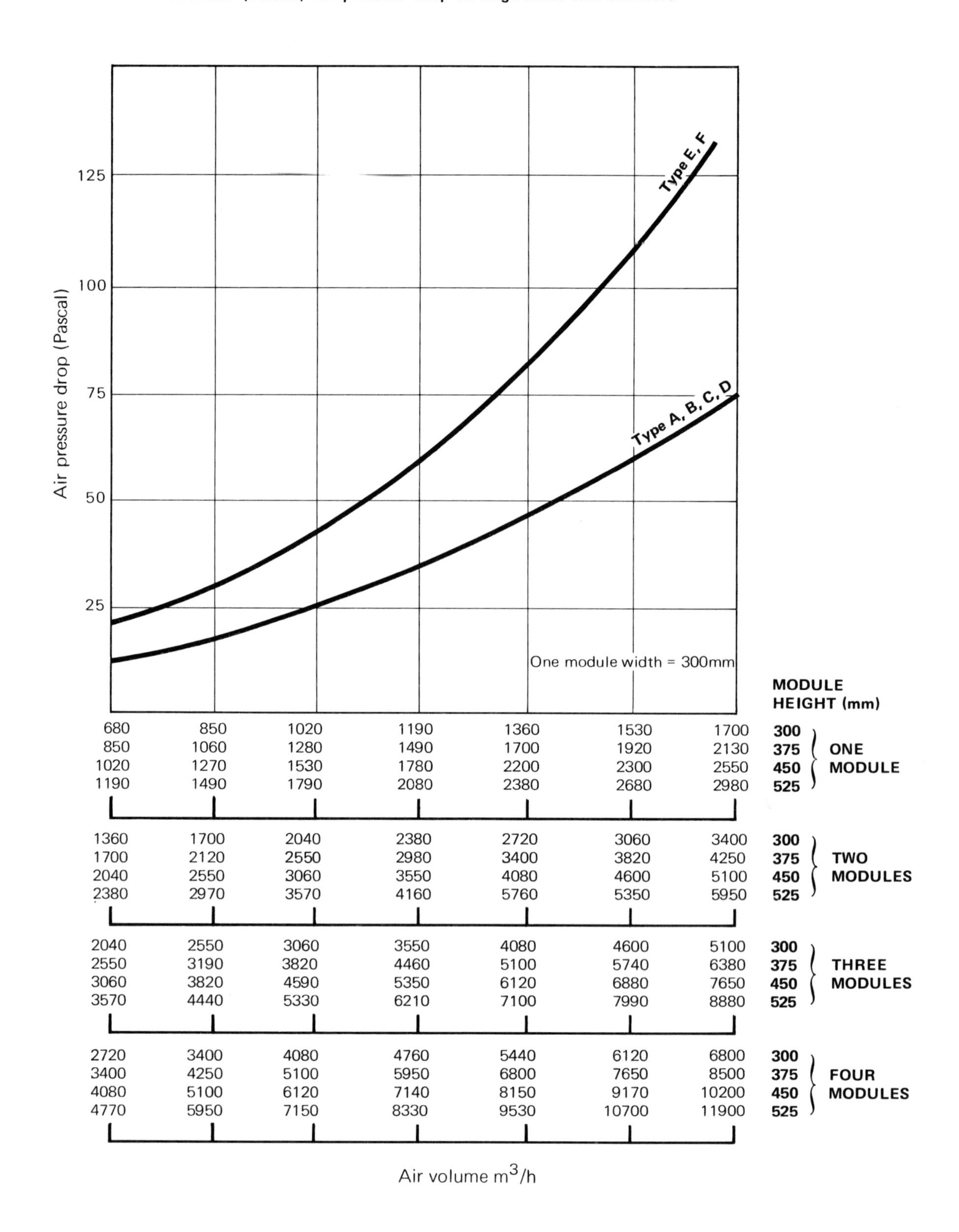

FIG 5.2 (Imperial) Air pressure drop through cross talk silencers

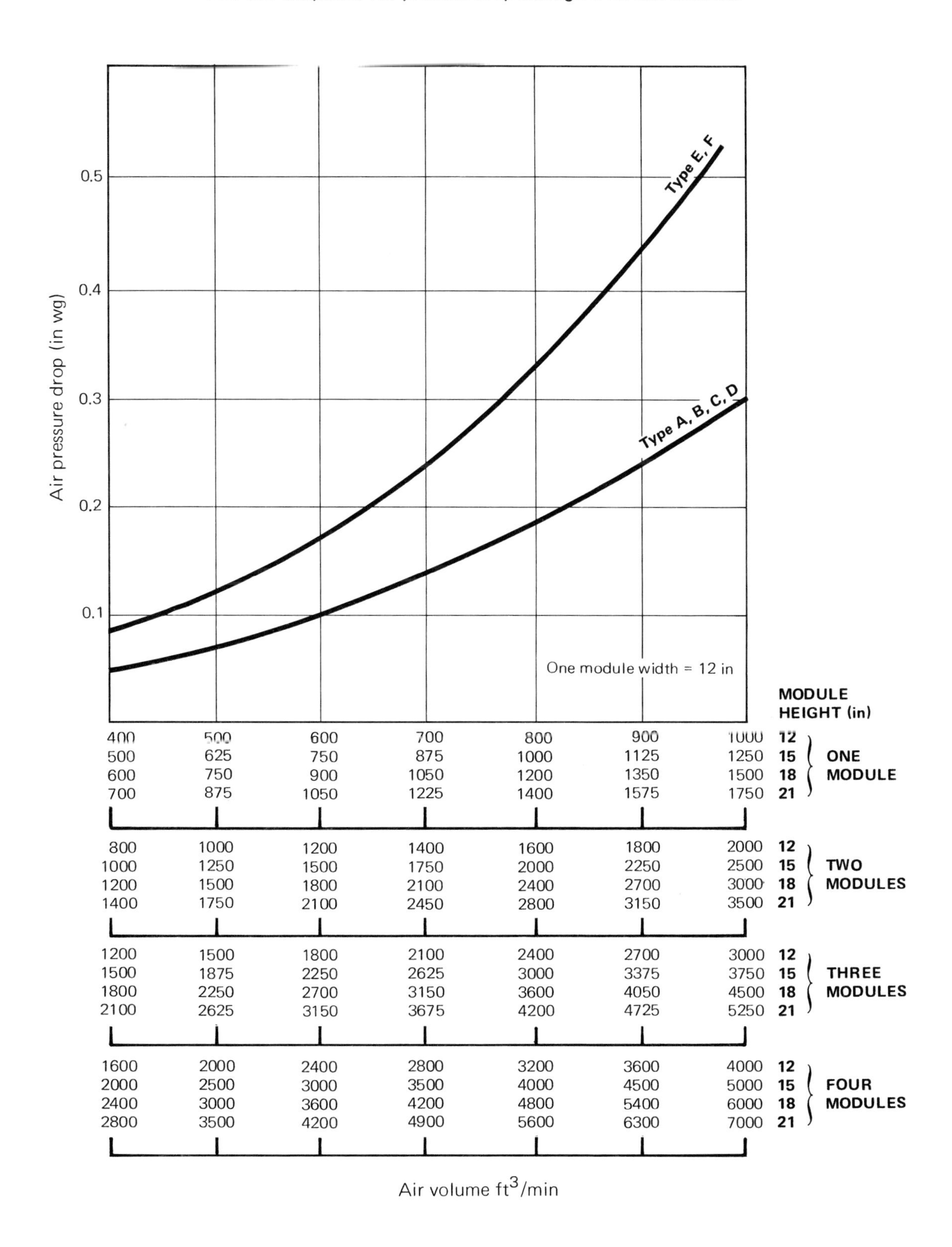

5.3 Location of silencers.

To ensure maximum effectiveness, silencers must be correctly located in the duct system. The location in general must be such that the breakout noise or flanking does not present problems. The optimum position of a silencer may vary from one installation to another, but the following points are usually applicable.

The silencer should be located as close to the fan as possible, particularly if the duct immediately downstream of the fan is over a critical area. This will reduce the in-duct power level and thereby ensure a minimum of breakout from that duct.

If the duct immediately downstream of the fan is over an area not sensitive to noise (e.g. a plant room) and the duct passes through a wall before it is over a relatively quieter area, the silencer may be placed in the partition as shown in Fig. 5.3. This will prevent the noise from the plant room entering the duct (i.e. breakin) reaching the adjacent quiet area.

Frequently, however, fire dampers are required in the plant room wall, and therefore the silencers may have to be located in a position other than as shown in Fig. 5.3. In such situations the silencer should be located on the room side of the fire damper. The silencer, and any duct between the silencer and the wall must be encased from the wall to the far end of the silencer by a suitable material.

The practice of locating silencers as shown in Fig. 5.4 should be avoided. Such a location is incapable of preventing noise breakout or breakin and reduces the effectiveness of the silencer.

The same criterion applies for the location of cross talk silencers: therefore, these too would be ideally situated in the partition wall between the two rooms.

FIG 5.3 Suitable location for a silencer

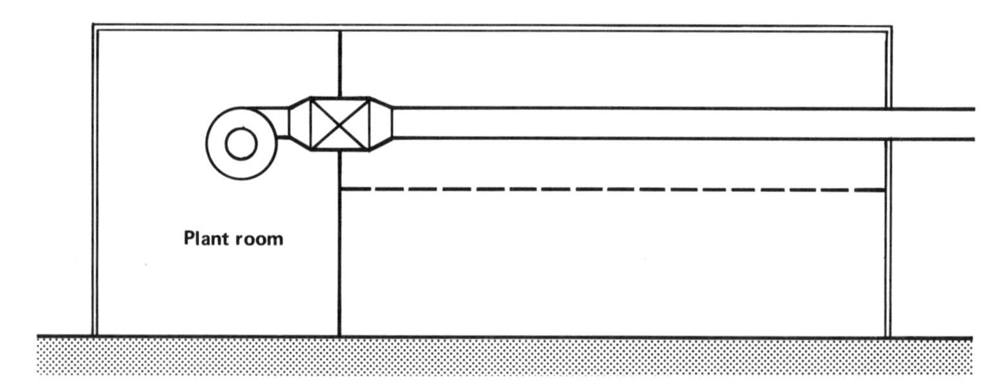

FIG 5.4 Unsuitable location for a silencer

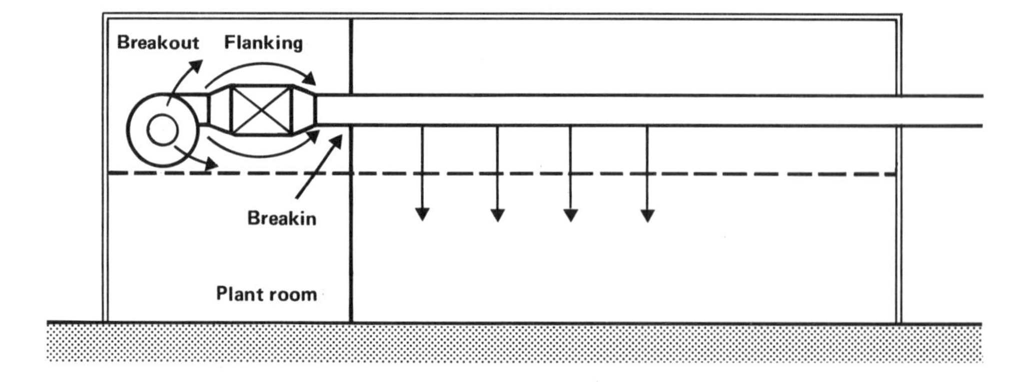

EXAMPLE:

Calculate a suitable cross talk silencer in a duct 450 mm x 200 mm (18 in x 8 in) handling approximately 1000 m^3/h (950 ft^3/min), interconnecting rooms 3 and 4 to the boardroom and director's office in the previous example (see layout, Fig. 2.1, pp. 13 and 15) maximum tolerable pressure drop 25 Pascal (0.1 in wg).

SOLUTION

Required NC level 35.

Duct dimensions = 450 mm x 200 mm (18 in x 8 in).

Group of silencer (from Tables 5.1 and 5.2) = *E* (overall width 225 mm (9 in), length 1.2 m (4 ft)).

One duct dimension i.e. 450 mm (18 in) fixes the number of modules i.e. two 225 mm (9 in).

Maximum flow for two 300 mm (12 in) modules is 1350 m^3/h (900 ft^3/min) (from Table 5.2).

As the required airflow exceeds this either the module height should be increased to 380 mm (15 in) or the number of modules increased to 3 (depending upon the duct size) to give the most suitable silencer dimensions.

For the given duct, i.e. 450 mm x 200 mm (18 in x 8 in), two modules of type E each 380 mm (15 in) high give an overall silencer dimension of 380 mm x 450 mm (15 in x 18 in) which appears quite suitable. Note: No silencer dimension must be smaller than a duct dimension. It is desirable, however, to keep at least one dimension the same while enlarging the other; this usually involves using expansion pieces, and it must be remembered that the cone angle of such an expansion piece should not exceed 30°.

In this situation, the air pressure drop = 20 Pascal (0.08 in wg) (from Fig. 5.2).

Location:
The silencer may be located in the branch KLM (preferably where it passes through the partition wall).

CHAPTER SIX

Noise to exterior

6.1 Introduction.

A complete analysis of noise to the exterior will normally include the evaluation and treatment of.

(i) Noise radiated from fan inlets and outlets to the atmosphere.
(ii) Noise radiated by the plant sited externally such as air cooled condensers and cooling towers.
(iii) Noise breaking out from plantroom walls and ceilings.

For the comprehensive treatment of item (iii), detailed calculations are necessary and an acoustic specialist should be consulted. For items (i) and (ii) however, the simple method described below will give usable solutions to the problem.

6.2 Background noise levels.

It is first necessary to establish whether noise breaking out or being generated externally could create unreasonable nuisance to people living or working close to the site. This is naturally dependent upon the type of neighbourhood and the time of day when the noise is generated. For example, night-time background noise levels drop considerably in comparison with day-time noise levels and therefore if plant operates for twenty four hours of the day, it will be necessary to adopt a more stringent standard than if it were working only during the daytime hours. Table 6.1 (page 71) gives an indication of the acceptable NC levels for various environments which, if exceeded, may cause annoyance. However, it should be noted that background noise levels are extremely variable between apparently similar areas and is is most important that, for a final detailed design, measurements are made of the existing background noise level. Once an assessment of the background noise level has been made, the starting point for the calculations is the acoustic power level of the source.

Fig. 6.1 Nomograph for calculating exterior noise levels

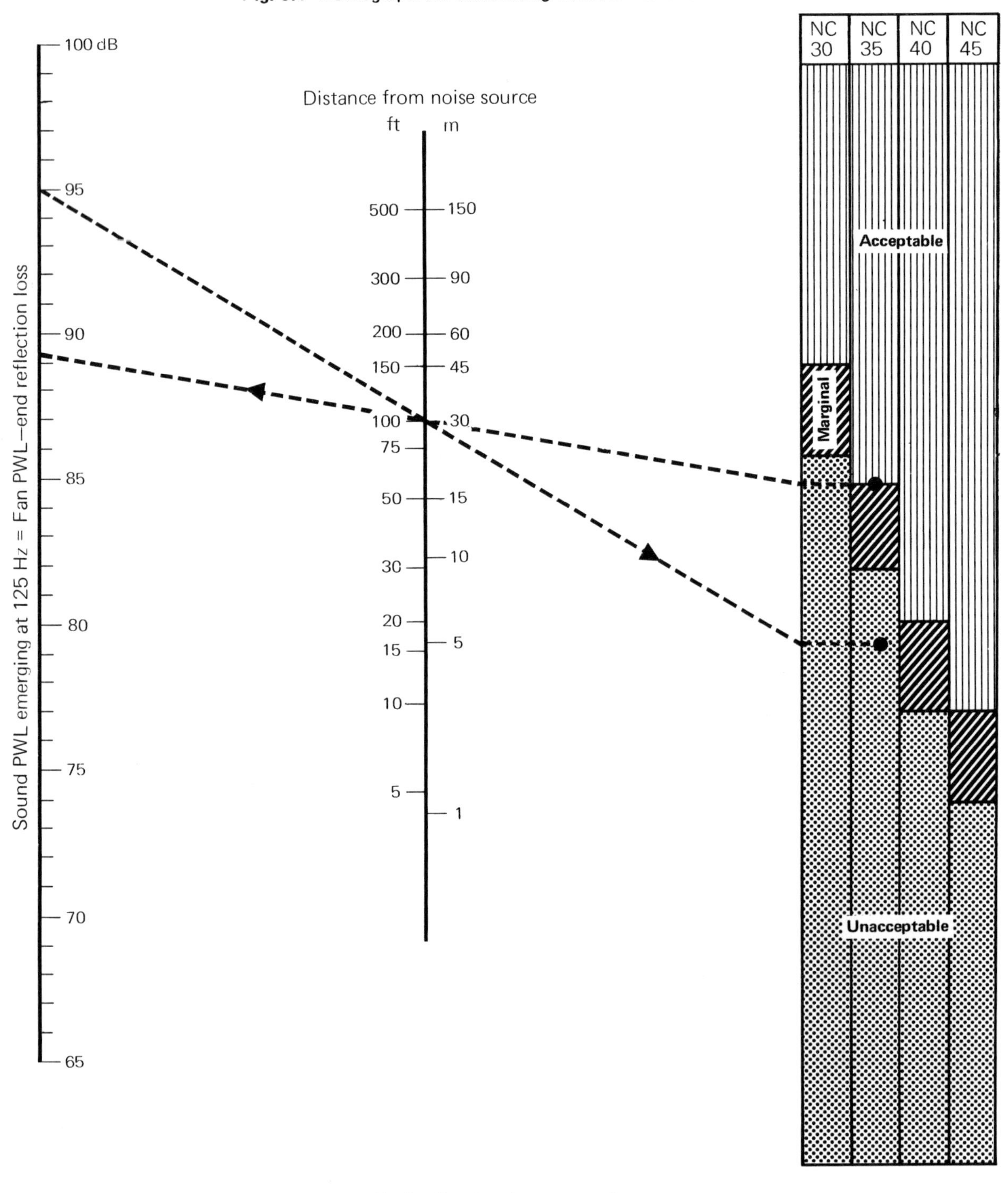

End reflection loss correction

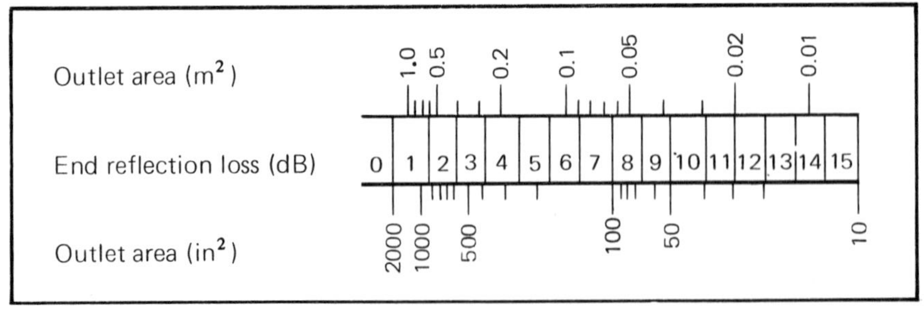

TABLE 6.1 Acceptable levels

Environments	*Background reference, NC*
Night-time rural; no nearby traffic of concern	15
Daytime rural; no nearby traffic of concern	25
Night-time suburban; no nearby traffic of concern	30
Daytime suburban; no nearby traffic of concern	35
Night-time urban; no nearby traffic of concern	30
Daytime urban; no nearby traffic of concern	40
Night-time business or commercial area	40
Daytime business or commercial area	45
Night-time industrial or manufacturing area	45

6.3 Noise from fan inlets and outlets.

For ventilation fans, the sound power level should have already been established during the duct silencing calculations.

For fans mounted in a duct (and in cooling towers) an end reflection loss factor must be subtracted from the in-duct power level. The resulting sound pressure levels in the neighbourhood will then be dependent on distance from the noise source. On the nomograph in Fig. 6.1, a straight line drawn from the emerging sound PWL through the distance scale will determine the acceptability of the installation to various NC criteria. Note, however, that obstructions, such as buildings, fences, or hilly terrain, occurring between the noise source and the neighbouring area concerned, in general reduce the noise transmitted along that particular path.

For all fans, whether used in ventilation or cooling plants the orientation of the outlet (or inlet) can alter the radiation of noise. The effect depends on the size of the outlet, its position (eg flush with a surface or in free space), and the angle between the outlet and the observer: in general small outlets will radiate equally in all directions, whereas large outlets have a strong directional characteristic on their axis. This means that, wherever possible, outlets and inlets should be sited so that they point at 45° or greater away from critical areas. A 3 to 6 dB increase may be expected (depending upon duct size and frequency) if the observer is directly opposite the inlet or outlet. (An allowance of 3dB for this is already included in Fig. 6.1.)

6.4 Noise from condensers and cooling towers.

For externally sited plant such as air cooled condensers and cooling towers, manufacturers' sound power level data may not be generally available. Some manufacturers do have a record of power levels produced by their items of plant but for those who do not have noise levels available, an approximate method of calculating sound power levels for typical cooling towers and condensers is given in figures A4.1 and A4.2.

The reduction of noise with distance and direction (see page 21) is calculated for both the inlet and outlet. The sound pressure level at the nearest noise sensitive point is then compared with the background noise level at that point. It may be assumed that the sound power level at the inlets and outlets of both cooling towers and condensers are equal except for axial fan-powered towers where the adjustments given in Fig. A4.2 should be made to the inlets. If it is found that a problem will exist, then the following possibilities should be examined to establish whether noise reduction is feasible.

(i) Re-positioning or re-orientation of equipment in open air.
(ii) Re-selection of equipment in open air.
(iii) Provision of silencers for equipment in open air or for fan inlets and outlets to atmosphere.

6.4.1 Re-positioning or re-orientation of equipment

For both cooling towers and condensers, re-positioning of equipment will generally only be of help if the plant can be sited either at a much greater distance from the receiver or behind some large obstruction (e.g. a building or solid fence but not trees) such that the line of sight between the receiver and the source is cut.

The orientation of the unit can make some difference to the sound pressure level transmitted to local residents. For cooling towers with axial fans the only parameter which can be varied with any degree of success is its position in the horizontal plane to ensure that a direct view of the outlet is not present. For those with centrifugal fans the re-orientation of the fan inlet in the horizontal plane will provide a considerable reduction in noise level at locations fairly close to the fan. In both cases these changes will make relatively little difference at more distant locations.

For condensers, only re-positioning at a much greater distance from the receiver can make any real difference to the noise transmission.

6.4.2 Re-selection of equipment

It is possible that a careful examination of the equipment requirements will allow a change of equipment, either by changing the type of plant or by using a different unit from the same range.

6.4.3 Provision of silencers

If all else fails and it becomes necessary to position a noisy item of plant where it is likely to cause a nuisance, the plant must be provided with silencers. It is extremely difficult to do this in the case of condensers, since the fans which are used in them are generally not capable of overcoming the resistance of the silencers. Therefore if low resistance silencing methods such as an enclosure with large free area silencers are used, the cost of such a method can become prohibitive in relation to the initial cost of the condensers. Under these conditions, it is recommended that the whole concept of using an air cooled condenser is reviewed.

In the case of cooling towers it is usually possible, given space, to provide adequate silencers.

In order to do this, first look at the type of fan used in the cooling tower. This may be either a centrifugal or an axial fan, mounted in a casing, and passing air over the heat exchange surface. Centrifugal fans are usually mounted on a common shaft and driven from one end by a standard motor with the air being drawn in at the bottom of one long side of the cooling tower and exhausted at the top of the cooling tower as shown in Fig. 6.2. Axial fans operate with air drawn in at the bottom of both sides, as shown in Fig. 6.3 and exhaust to atmosphere directly upwards. In both cases, fans are usually sized so that they produce the required amount of air flow against a small additional resistance. This means that if silencing is provided, great care must be taken to ensure that the free area of the silencers is generous enough to keep that additional resistance small. The tower manufacturers will be able to advise on the maximum allowable resistance.

For normal fan inlets and outlets to atmosphere, attenuators should be selected in the same way as those on the room-side of each fan.

How much silencing?

Using the nomograph on page 70 it is possible to arrive at the sound pressure level created by a noise source at a given distance. Any excess over the background noise level which has to be reduced by silencing can be obtained by working the nomograph backwards. This will give the insertion loss of a silencer in the 125 Hz band, and a silencer chosen on this figure alone will in general meet the requirements of all other frequency bands.

FIG. 6.2 Silencing of a centrifugal fan cooling tower

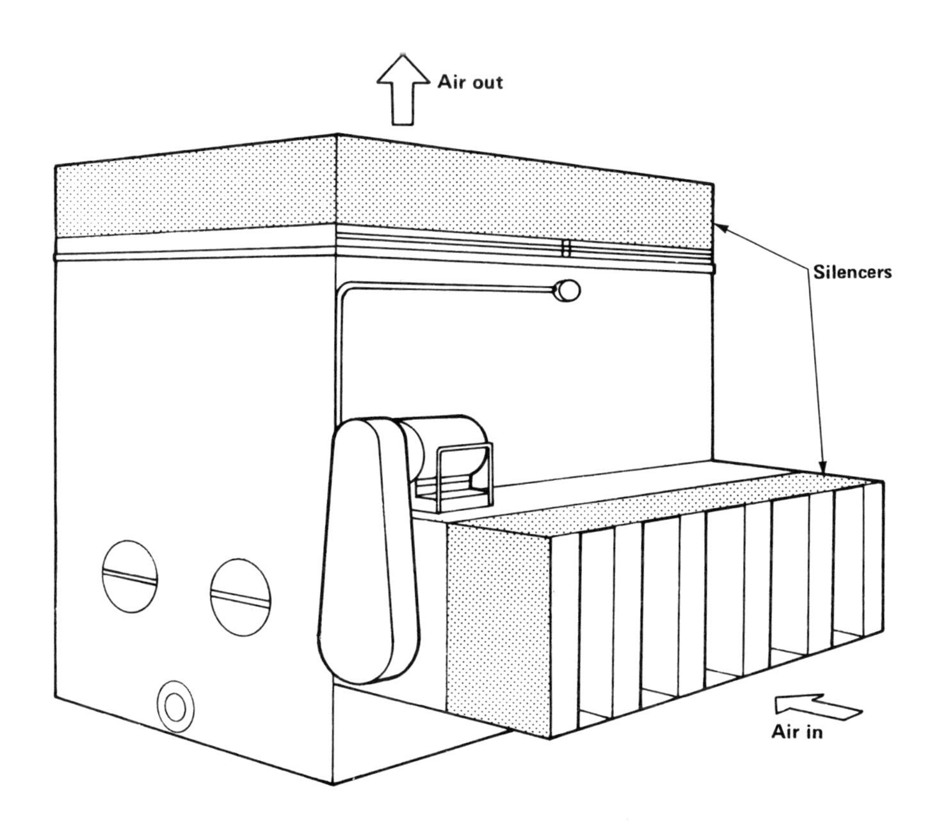

FIG. 6.3 Silencing of an axial fan cooling tower

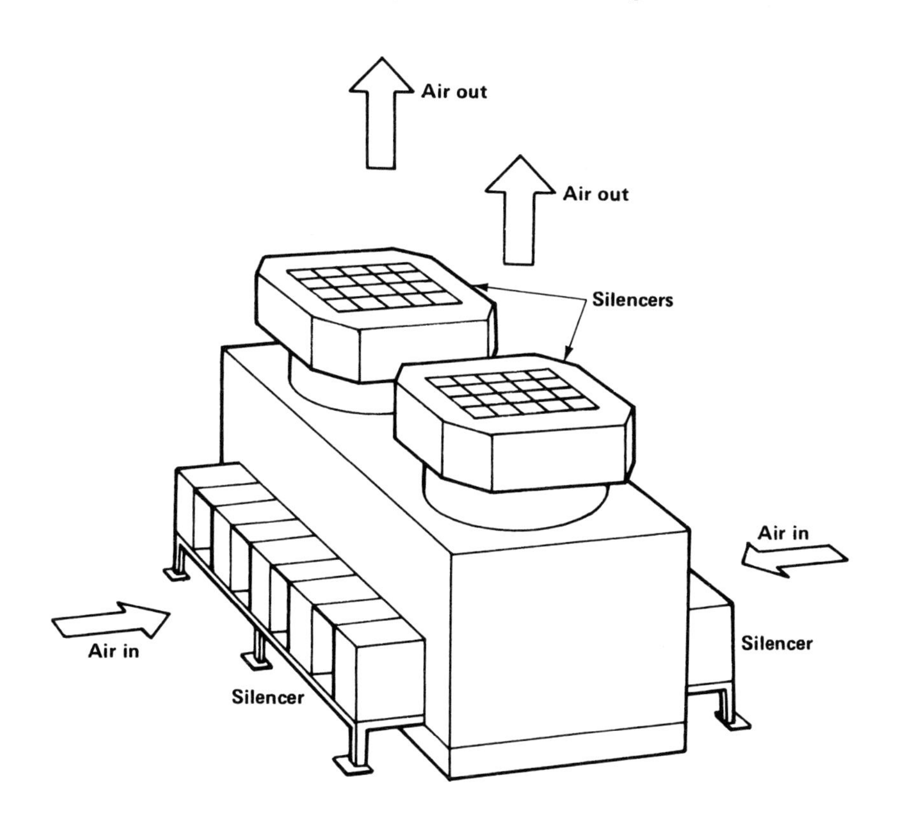

EXAMPLE:

An exhaust fan having a PWL at 125 Hz of 96 dB discharges air into the atmosphere twenty four hours a day. It is mounted in a short length of duct of cross section 1200 mm x 600 mm (4 ft x 2 ft). Estimate whether this will cause annoyance to a residential neighbourhood, the nearest house being 30 m (100 ft) away, where the background noise level at night drops to NC 35.

If annoyance is likely, calculate the 125 Hz band insertion loss of a silencer which would make the noise level in the neighbourhood acceptable.

The air volume is 25,000 m^3/h (15000 ft^3/min) and the maximum permissible pressure drop is 75 Pascal (0.3 in wg).

SOLUTION

Fan PWL at 125 Hz = 96 dB
End reflection loss for duct termination 1200 mm x 600 mm (4ft x 2ft) = 1 dB
i.e. 0.72 m^2 (1152 in^2)
$\therefore$ Fan PWL emerging = 95 dB

From the distance loss nomograph Fig. 6.1 on page 70 draw a line from the left scale at 95 dB through the middle scale at 30 m (100 ft). It can be seen that the extended line enters the NC 30 column at the unacceptable level.

For assessing the amount of silencing required to reduce the level to one which is acceptable, simply work the nomograph backwards. Draw a line from the acceptable level in the NC 35 column through the 30 m (100 ft) scale. It can be seen that the line terminates at the 89 dB level on the left-hand scale.

The required silencing is therefore the unacceptable level	95 dB
minus acceptable level	89 dB
equals	6 dB

From the silencer selection chart Fig. A3.1 page 85 it can be seen that to achieve 6 dB silencing at 125 Hz a 0.9 m (36 in) long attenuator is needed.

From Figure A3.2 the maximum passage velocity allowable in order not to exceed a 75 Pascal (0.3 in wg) pressure drop is 15 m/s (3000 ft/min).

The net free area is determined thus:

$$NFA = \frac{\text{Air volume}}{\text{Passage velocity}}$$

In this example

(i) (metric) $$NFA = \frac{8800\ m^3/h}{15 m/s \times 3600} = 0.5 m^2$$

Therefore from Table A3.1—a 1050mm x 1050mm silencer is selected.

(ii) (imperial) $$NFA = \frac{15000\ ft^3/min}{3000\ ft/min} = 5\ ft^2$$

Therefore from Table A3.1—a 42in x 42in silencer is selected.

CHAPTER SEVEN

Installation and commissioning

All the effort in designing the system with care and thought can be wasted if the equipment is not installed with an equal amount of care. Apply the following nine-point check plan:

(1) When the plant room has been completed, check to ensure that no holes or gaps exist in it. If these are found, they should be filled with concrete or a similar impervious material. All holes through which ducts, pipes, etc. pass should be filled with a suitable material. Note that fibreglass, polystyrene or newspaper, etc. are not suitable materials.

(2) Adequate care must be taken to ensure that the anti-vibration mountings used with compressors, pumps, etc. are not short circuited by rigidly fixed pipework, electrical conduit, or builder's rubble.

(3) Loose dampers in the ducts must be avoided, as these are a common source of noise. If found loose, they must be properly tightened. Check that the damper indicators indicate the damper position correctly.

(4) Large unstiffened duct walls must be avoided as this may result in drumming. Where this is so, adequate stiffeners of substantial rigidity should be provided. Where lagging of ducts is required, care must be taken to lag the entire surface. It is quite common to find the top of the duct, close to the ceiling, left unlagged.

(5) Check for misaligned duct joints, and rectify.

(6) Check for leaks in ducts. All holes required for pressure or velocity measuring apparatus or any other purpose should be sealed after tests have been completed.

(7) Check that lighting and ceilings are not suspended from ducts.

(8) Check that the ventilation system is properly balanced.

(9) Installation of ductwork must be strictly supervised. In one airconditioning system, plastic coffee cups and lunch paper bags thrown in the duct during installation caused considerable noise. Clearing out such refuse can be difficult once the system is in operation.

APPENDIX 1

Sound reduction indices (transmission loss) of common structures

Representative values of airborne sound reduction index for some common structures adapted from 'Airborne Sound Insulation of Partitions' (HMSO 1966) by permission of the Controller of Her Majesty's Stationery Office

Partition construction	Thickness (mm)	Superficial weight (kg/m^2)	Frequency (Hz)							
			63	125	250	500	1000	2000	4000	8000
Panels of sheet materials										
1.5 mm lead sheet	1.5	17	22	28	32	33	32	32	33	36
3 mm lead sheet	3	34	25	30	31	27	38	44	33	38
20 g aluminium sheet, stiffened	0.9	2.5	8	11	10	10	18	23	25	30
22 g galvanised sheet steel	0.55	6	3	8	14	20	23	26	27	35
20 g galvanised sheet steel	0.9	7	3	8	14	20	26	32	38	40
18 g galvanised sheet steel	1.2	10	8	13	20	24	29	33	39	44
16 g galvanised sheet steel	1.6	13	9	14	21	27	32	37	43	42
18 g fluted steel panels stiffened at edges, joints sealed	1.2	39	25	30	20	22	30	28	31	31
Corrugated asbestos sheet stiffened and sealed	6	10	20	25	30	33	33	38	39	42
Chipboard sheets on wood framework	19	11	14	17	18	25	30	26	32	38
Fibreboard sheets on wood framework	12	4	10	12	16	20	24	30	31	36
Plasterboard sheets on wood framework	9	7	9	15	20	24	29	32	35	38
Plywood sheets on wood framework	6	3.5	6	9	13	16	21	27	29	33
Hardwood (mahogany) panels	50	25	15	19	23	25	30	37	42	46
Woodwool slabs unplastered	25	19	0	0	2	6	6	8	8	10
Woodwool slabs plastered (12 mm on each face)	50	75	18	23	27	30	32	36	39	43
Panels of sandwich construction										
1.5 mm lead between two sheets of 5 mm plywood	11.5	25	19	26	30	34	38	42	44	47
9 mm asbestos board between two sheets of 18 g steel	12	37	16	22	27	31	27	37	44	48
'Stramit' compressed straw between two sheets of 3 mm hardboard	56	25	15	22	23	27	27	35	35	38
Single masonry walls										
Single leaf brick, plastered both sides	125	240	30	36	37	40	46	54	57	59
" "	255	480	34	41	45	48	56	65	69	72
" "	360	720	36	44	43	49	57	66	70	72

Representative values of airborne scund reduction index for some common structures (contd.)

Partition construction	Thickness (mm)	Superficial weight (kg/m^2)	Frequency (Hz)							
			63	125	250	500	1000	2000	4000	8000
Solid breeze or clinker blocks, plastered (12 mm both sides)	125	145	20	27	33	40	50	57	56	59
Solid breeze or clinker blocks, unplastered	75	85	12	17	18	20	24	30	38	43
Hollow cinder concrete blocks, painted (cement base paint)	100	75	22	30	34	40	50	50	52	53
Hollow cinder concrete blocks, unpainted	100	75	22	27	32	37	40	41	45	48
'Thermalite' blocks	100	125	20	27	31	39	45	53	38	62
Glass bricks	200	510	25	30	35	40	49	49	43	45
Double masonry walls										
280 mm brick, 56 mm cavity, strip ties, outer faces plastered 12 mm	300	380	28	34	34	40	56	73	76	78
280 mm brick, 56 mm cavity, expanded metal ties, outer faces plastered, 12 mm	300	380	27	27	43	55	66	77	85	85
Stud partitions										
50 mm x 100 mm studs, 12 mm insulating board both sides	125	19	12	16	22	28	38	50	52	55
50 mm x 100 mm studs, 9 mm plaster board and 12 mm plaster coat both sides	142	60	20	25	28	34	47	39	50	56
Single glazed windows										
Single glass in heavy frame	6	15	17	11	24	28	32	27	35	39
	8	20	17	18	25	31	32	28	36	39
	9	22.5	18	22	26	31	30	32	39	43
	16	40	20	25	28	33	30	38	45	48
	25	62.5	25	27	31	30	33	43	48	53
Double glazed windows										
4 mm panes, 7 mm cavity	12	15	15	22	16	20	29	31	27	30
9 mm glass panes in separate frames, 50 mm cavity	62	34	18	25	29	34	41	45	53	50
6 mm glass panes in separate frames, 100 mm cavity	112	34	20	28	30	38	45	45	53	50
6 mm glass panes in separate frames, 188 mm cavity	200	34	25	30	35	41	48	50	56	56
6 mm glass panes in separate frames, 188 mm cavity with absorbent blanket in reveals	200	34	26	33	39	42	48	50	57	60
6 mm and 9 mm panes in separate frames, 200 mm cavity, absorbent blanket in reveals	215	42	27	36	45	58	59	55	66	70

Representative values of airborne sound reduction index for some common structures (contd.)

Partition construction	*Thickness (mm)*	*Superficial weight (kg/m^2)*	*Frequency (Hz)* 63	125	250	500	1000	2000	4000	8000
Doors										
Flush panel, hollow core, normal cracks as usually hung	43	9	9	12	13	14	16	18	24	26
Solid hardwood, normal cracks as usually hung	43	28	13	17	21	26	29	31	34	32
Typical proprietary 'acoustic' door, double heavy sheet steel skin, absorbent in airspace, special furniture and seals in heavy steel frame	100	–	37	35	39	44	49	54	57	60
Floors										
T & G boards, joints sealed	21	13	17	21	18	22	24	30	33	63
T & G boards, 12 mm plasterboard ceiling under, with 3 mm plaster skim coat	235	31	15	18	25	37	39	45	45	48
As above with boards 'floating' on glass wool mat	240	35	20	25	33	38	45	56	61	64
Concrete, reinforced	100	230	32	37	36	45	52	59	62	63
" "	200	460	36	42	41	50	57	60	65	70
" "	300	690	37	40	45	52	59	63	67	72
126 mm reinforced concrete with 'floating' screed	190	420	35	38	43	48	54	61	63	67

APPENDIX 2

Absorption coefficients of common internal finishes

Approximate values of absorption coefficient for some common internal finishes adapted from 'Sound Absorbing Materials' (HMSO 1960) by permission of the Controller of Her Majesty's Stationery Office.

Material	*Thickness (including any airspace) (mm)*	*Frequency (Hz)*							
		63	125	250	500	1000	2000	4000	8000
Normal wall finishes									
Brickwork	–	0.05	0.05	0.04	0.02	0.04	0.05	0.05	0.05
Breeze or cinder block	–	0.10	0.20	0.45	0.60	0.40	0.45	0.40	0.40
Concrete	–	0.01	0.01	0.01	0.02	0.02	0.02	0.03	0.03
Up to 4 mm thick glass pane about 1 metre square	4	0.25	0.35	0.25	0.20	0.10	0.05	0.05	0.05
6 mm plate glass about 1 metre square	6	0.08	0.15	0.06	0.04	0.03	0.02	0.02	0.02
Marble or glazed tile	–	0.05	0.05	0.05	0.05	0.05	0.05	0.05	0.05
Plaster on solid wall	12	0.04	0.04	0.05	0.06	0.08	0.04	0.06	0.05
Water (e.g. swimming pool)	–	0.01	0.01	0.01	0.01	0.01	0.02	0.02	0.02
Wall or ceiling treatments									
Curtains hung in folds against solid wall	–	0.05	0.05	0.15	0.35	0.40	0.50	0.50	0.40
'Acoustic' plaster (typical values)	12	0.05	0.10	0.15	0.20	0.25	0.30	0.35	0.35
Sprayed asbestos, direct on wall or ceiling	25	0.05	0.10	0.30	0.65	0.85	0.85	0.80	0.75
" " on expanded metal with 75 mm air gap	100	0.20	0.30	0.40	0.65	0.80	0.75	0.75	0.70
Glass or rockwool blanket, typical values for medium density material	25	0.05	0.10	0.35	0.60	0.70	0.75	0.80	0.75
	50	0.10	0.20	0.45	0.65	0.75	0.80	0.80	0.80
	100	0.25	0.45	0.75	0.80	0.85	0.85	0.90	0.85
	150	0.35	0.55	0.90	0.90	0.85	0.90	0.95	0.95
Expanded polyurethane foam (open cell)	25	0.10	0.15	0.30	0.60	0.75	0.85	0.90	0.90
	50	0.15	0.25	0.50	0.85	0.95	0.90	0.90	0.90
	100	0.30	0.50	0.70	0.95	1.00	1.00	1.00	1.00
9 mm plasterboard on battens at 0.5 m centres, 18 mm airspace filled with glass wool	27	0.25	0.30	0.20	0.15	0.05	0.05	0.05	0.05
5 mm plywood on battens at 1 m centres, 50 mm airspace filled with glass wool	55	0.30	0.40	0.35	0.20	0.15	0.05	0.05	0.05
12 mm plywood on battens at 1 m centres, 59 mm airspace filled with glass wool	71	0.25	0.30	0.20	0.15	0.10	0.15	0.10	0.05
3 mm hardboard with roofing felt stuck to back over 50 mm airspace	53	0.50	0.90	0.45	0.25	0.15	0.10	0.10	0.05
Suspended plaster or plasterboard ceiling (large airspace)	–	0.20	0.20	0.15	0.10	0.05	0.05	0.05	0.05
Fibreboard on solid backing	12	0.05	0.05	0.10	0.15	0.25	0.30	0.30	0.25

Approximate values of absorption coefficient for some common internal finishes (contd)

Material	Thickness (including any airspace) (mm)	Frequency (Hz) 63	125	250	500	1000	2000	4000	8000
Floor coverings									
Composition flooring	–	0.05	0.05	0.05	0.05	0.05	0.05	0.05	0.05
Haircord carpet on felt underlay	6	0.05	0.05	0.05	0.10	0.20	0.45	0.65	0.65
Medium pile carpet on sponge rubber underlay	10	0.05	0.05	0.10	0.30	0.50	0.65	0.70	0.65
Thick pile carpet on sponge rubber underlay	15	0.05	0.15	0.25	0.50	0.60	0.70	0.70	0.65
Rubber floor tiles	6	0.05	0.05	0.05	0.10	0.10	0.05	0.05	0.05
Proprietary acoustic tiles and boards									
(Note that performance varies according to individual construction and method of fixing. Always obtain exact figures from manufacturer. The figures shown indicate the likely range of performance)									
Fixed direct on wall or ceiling, or with small airspace minimum	12 to 75	0.05	0.10	0.25	0.50	0.60	0.60	0.45	0.45
maximum	12 to 75	0.15	0.20	0.60	0.80	0.85	0.80	0.75	0.75
In the form of suspended ceiling minimum	–	0.15	0.30	0.40	0.50	0.65	0.75	0.70	0.65
maximum	–	0.30	0.50	0.60	0.90	0.90	0.85	0.80	0.75
Room contents									
(Figures shown are total value of *Sa* in m^2 units)									
Audience per person in fully upholstered seat	–	0.15	0.20	0.40	0.45	0.45	0.50	0.45	0.40
Audience per person in wood or padded seat	–	0.10	0.15	0.25	0.40	0.40	0.45	0.40	0.35
Unoccupied seat fully upholstered	–	0.05	0.10	0.20	0.30	0.30	0.30	0.35	0.30
Unoccupied seat wood or padded	–	0.02	0.03	0.05	0.05	0.10	0.15	0.10	0.10

APPENDIX 3

Silencer selection data

FIG A3.1 Silencer length calculation

From this chart select a suitable length of silencer for the required insertion loss.

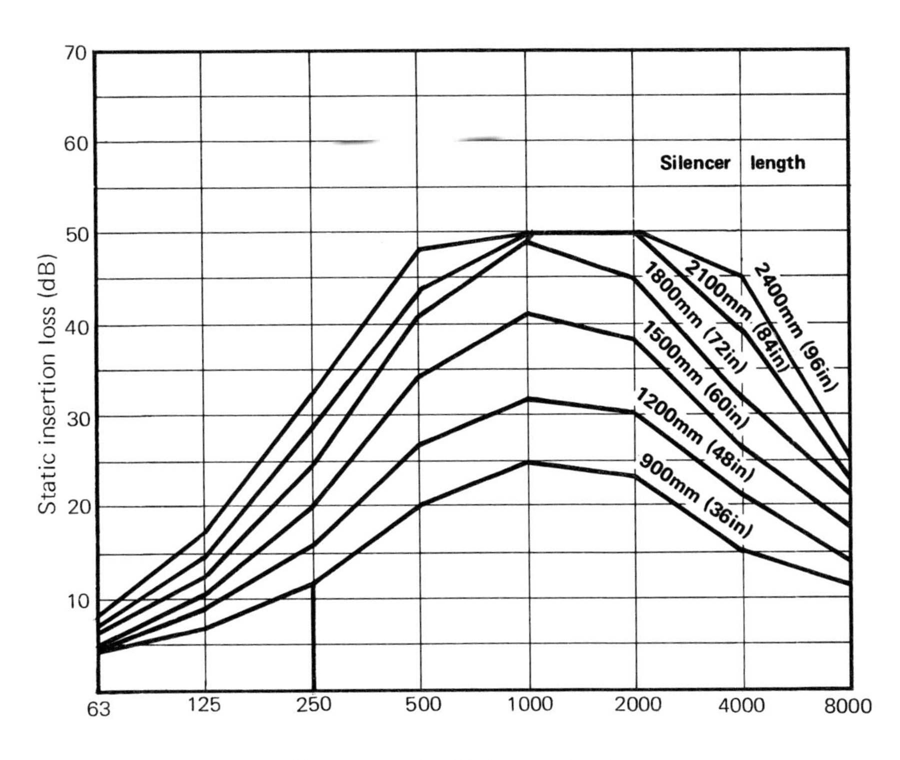

Based on data supplied by Sound Attenuators Ltd

FIG A3.2 Passage air velocity calculation

From this chart, knowing the allowable pressure drop, calculate the passage air velocity. Then determine nett free area.

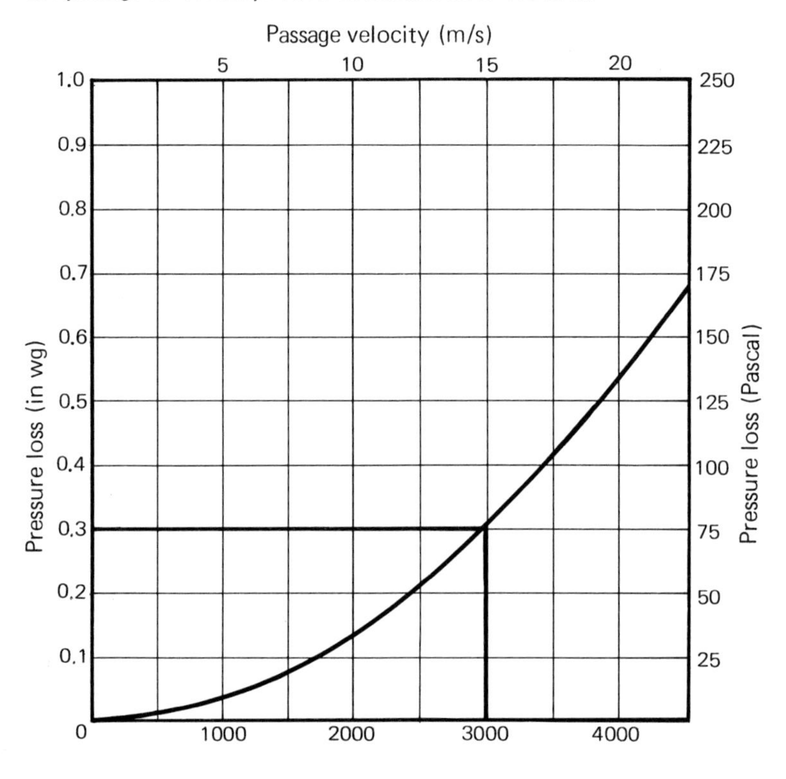

Based on data supplied by Sound Attenuators Ltd

TABLE A3.1 Silencer dimension calculations

Knowing the nett free area, determine suitable cross section dimensions using this table

Nett free area m^2	*Suitable dimensions*				*Nett free area* ft^2
	Metric		*Imperial*		
	Width *mm*	*Height* *mm*	*Width* *in*	*Height* *in*	
0.05	350	300	14	12	0.5
0.070	350	450	14	18	0.75
0.100	350	600	14	24	1.00
0.140	700	450	28	18	1.50
0.185	700	600	28	24	2.00
0.232	700	750	28	30	2.50
0.280	700	900	28	36	3.00
0.325	700	1050	28	42	3.50
0.280	1050	600	42	24	3.00
0.350	1050	750	42	30	3.75
0.420	1050	900	42	36	4.50
0.490	1050	1050	42	42	5.25
0.560	1050	1200	42	48	6.00
0.630	1050	1350	42	54	6.75
0.700	1050	1500	42	60	7.50
0.465	1400	750	56	30	5.00
0.560	1400	900	56	36	6.00
0.650	1400	1050	56	42	7.00
0.740	1400	1200	56	48	8.00
0.840	1400	1350	56	54	9.00
0.930	1400	1500	56	60	10.00
0.770	1750	900	70	36	7.50
0.813	1750	1050	70	42	8.75
0.930	1750	1200	70	48	10.00
1.045	1750	1350	70	54	11.25
1.161	1750	1500	70	60	12.50

Based on data supplied by Sound Attenuators Ltd.

APPENDIX 4

Sound power levels of cooling towers and air cooled condensers

FIG A4.1 PWL of outdoor sited fan equipment (e.g. aircooled condensers, cooling towers etc.)

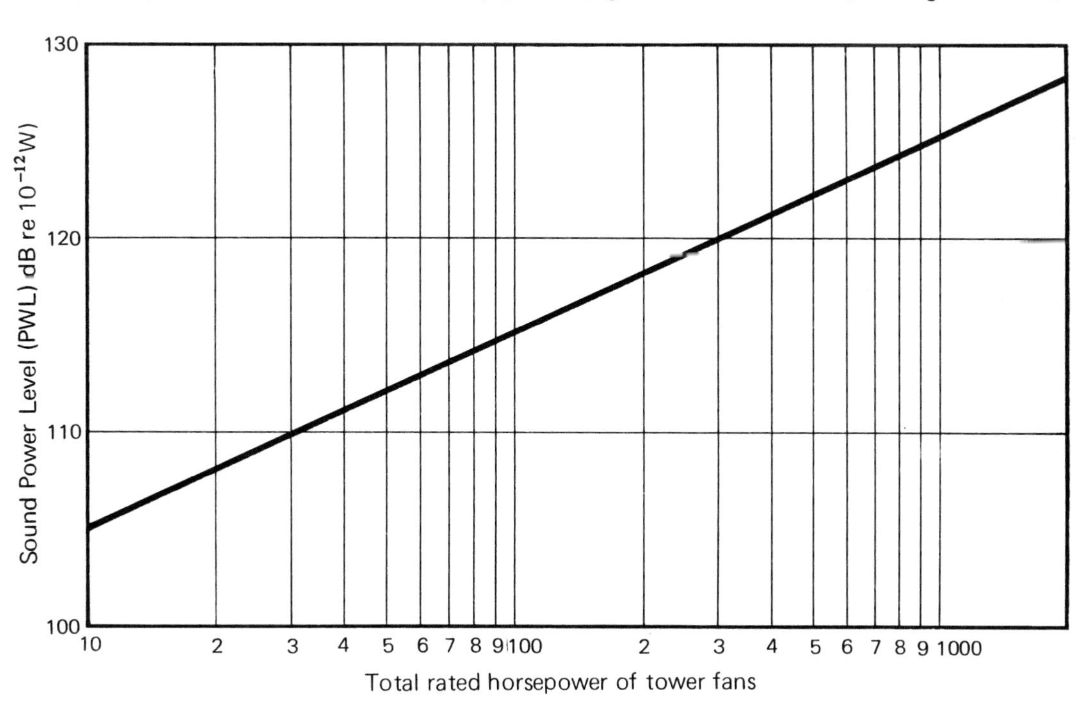

FIG A4.2 Spectrum corrections to be subtracted from values in Fig A4.1

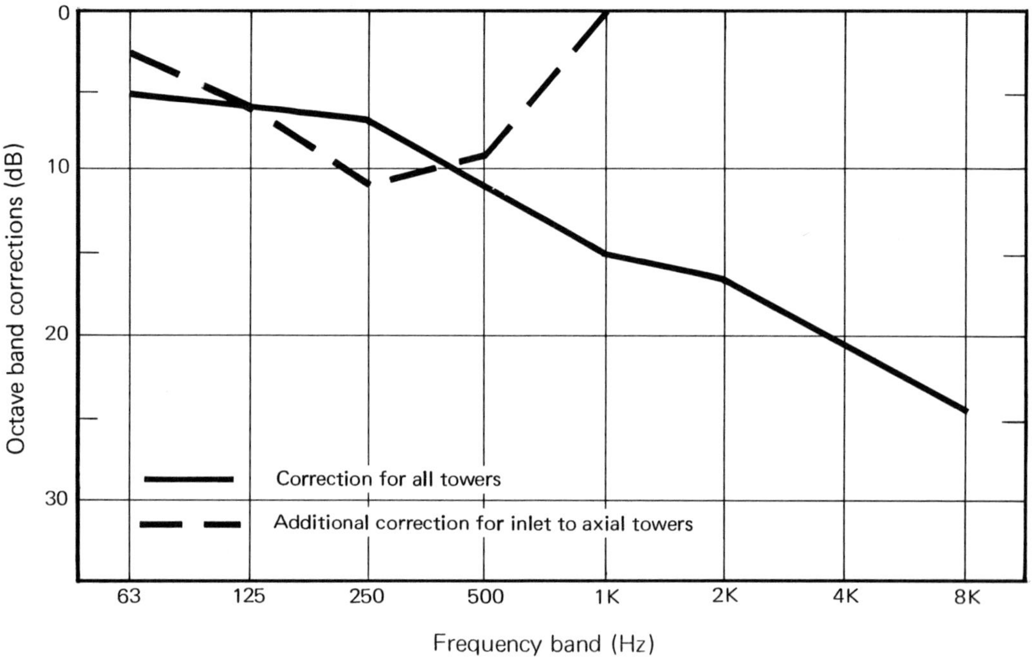

APPENDIX 5

Computer programs for noise analysis

A5.1 Introduction.

The calculations outlined in chapters 2, 3 and 4 for the acoustic analysis of ventilation duct systems can also be performed by a suite of computer programs devised by Atkins Computing Services and entitled VSAAP (Ventilation System Acoustic Analysis Programs).

Listed below are brief descriptions of the programs currently available. It is likely that additional programs will be added to the suite and alterations made to the existing programs. Users are therefore advised to obtain up to date information directly from Atkins Computing Services, Woodcote Grove, Ashley Road, Epsom, Surrey.

A5.2 VGN2. (velocity generated noise in duct bends).

This program calculates the PWL of velocity generated noise in duct bends of rectangular cross section whose aspect ratio does not exceed 2. The effect of introducing circular arc turning vanes may also be calculated. The technique and data used are as described in the ASHRAE Handbook (1973 Systems volume, ch. 35).

There are two versions of this program. The conversational version tells the user the exact data it requires, and asks for it as it needs it. When, however, a lot of bends in a large installation have to be analysed, the conversational version can be too slow, and the second version may be used. This takes the data in bulk for all the elbows in the system and prints out the spectra in rapid succession.

A5.3 VGN3. (velocity generated noise in branch takeoffs).

This program calculates the velocity generated noise at branch takeoffs in circular or rectangular cross-section ducts. The program is capable of analysing the effect of both sharp, and radiused takeoffs. The technique and data are described in ASHRAE Handbook (1973 Systems volume, ch. 35).

A5.4 LVAA. (low velocity air-conditioning duct acoustic analysis).

This program analyses a complete ventilation duct branch from the fan or air handling unit right through to the room outlet, and produces the resultant sound pressure level spectrum in the given room. Standard methods are used, as can be found in IHVE/ASHRAE guide books, and in addition provision is made for the inclusion of the effect of localised noise breakin or noise generation due to high velocity in bends and branches.

In addition to computing the grille borne noise, the program also carries out a duct breakout analysis and in both cases calculates the silencing required to satisfy a given maximum noise criterion.

A5.5 HVAA. (high velocity air-conditioning acoustic analysis).

This program calculates the "in-duct" and "breakout" sound power levels at each stage of a high velocity duct system, to assist the designer in selecting and positioning silencing equipment.

The program carries out a downstream analysis (away from the fan) and an up-stream analysis (towards the fan). The resultant noise level at any stage is calculated by combining logarithmically the velocity generated noise and the duct-borne noise of the fan at the stage considered. Allowance is made for natural attenuation of the fan noise by the ducting. The analysis is necessary in both directions because velocity generated noise is transmitted both upstream and downstream of its point of generation.

The calculation of duct "breakout" level is a conversational feature of the program and is an optional facility for users requiring this information.

The program has primarily been designed to analyse a main duct riser. Branch runs and sub-branch runs can be analysed by further runs of the program as required, using the "on-going" power level entering the branch as the starting power level. For this purpose, the "on-going" power level is calculated at each branch for each run of HVAA.

APPENDIX 6

Fundamentals and terminology

A6.1 Sounds, frequency and magnitude.

The following are explanations of the acoustic terms used in this book. They are not intended to be other than simplified descriptions applicable to this text. A more comprehensive list of definitions is to be found in British Standard 661 : 1969.

Noise and sound

Vibrating sources in contact with air create pressure fluctuations which appear to the ear as sound. Noise can be defined as that sound which is unpleasant or perhaps more broadly as that sound which is unwanted by the recipient.

Frequency

The number of pressure fluctuations per second governs the pitch of sound and is called the frequency. It is measured in cycles per second (cps) or hertz (Hz). The human ear responds to the frequencies approximately between 25 Hz and 15 000 Hz. Any range having an upper value twice that of the lower value is called an octave band.

Magnitude

The magnitude of sound may be expressed in terms of either sound power, sound intensity or sound pressure:

Sound power is the total rate of acoustical energy released by the sound source and is usually expressed in watts.

Sound intensity is the mean rate of acoustic energy flowing through unit area, normal to the direction of propagation. Logically it is expressed in watts per square centimetre.

Sound pressure is the value of the pressure variation from the ambient at a point within the sound field, usually expressed in either microbars, dynes per square centimetre or Newtons per square metre (1 N/m^2 = 10 $dyne/cm^2$ = 10^{-2} mb = 10 μb).

It will be appreciated that in respect of the given sound, the sound power will be constant, but the sound intensity and pressure will be dependent upon the conditions under which they are measured.

A6.2 The bel and decibel.

The unit bel (i.e. 10 dB) is borrowed from the field of telecommunication engineering, and is a dimensionless unit expressing the logarithmic ratio of two quantities proportional to power. However, the quantities sound power, intensity and sound pressure vary over a very wide range (as can be seen from Table A6.1); such large numbers are therefore clumsy to handle in calculations or when describing a sound source. The use of a logarithmic scale conveniently helps to overcome this problem.

As an example a sound power of 100 000 W may be written more compactly as 10^5 W. If we express this as a ratio, relative to say 1 W, this becomes 10^5. If we take the logarithmic ratio which defines the bel, this becomes even more compact, i.e. 5 bels or 50 dB.

$$\text{Therefore sound power level in dB} = 10 \log_{10} \frac{\text{sound power}}{\text{reference power}} .$$

The reference power in the above example was taken as 1 W. The reference generally accepted and arrived at by international agreement is 10^{-12} W, although a reference of 10^{-13} W has been very widely used in the USA. The level in decibels relative to the former reference can be arrived at by subtracting 10 dB from the level relative to the latter reference.

TABLE A6.1 Linear, exponential and decibel (logarithmic) scales for sound power

Radiated sound power (W)		*Sound power level (dB)*		
Usual notation	*Expressed as powers*	*Relative to 1 W*	*Relative to 10^{-13} W*	*Relative to 10^{-12} W*
100 000	10^5	50	180	170
10 000	10^4	40	170	160
1 000	10^3	30	160	150
100	10^2	20	150	140
10	10^1	10	140	130
1	1	0	130	120
0.1	10^{-1}	–10	120	110
0.01	10^{-2}	–20	110	100
0.001	10^{-3}	–30	100	90
0.000 1	10^{-4}	–40	90	80
0.000 01	10^{-5}	–50	80	70
0.000 001	10^{-6}	–60	70	60
0.000 000 1	10^{-7}	–70	60	50
0.000 000 01	10^{-8}	–80	50	40
0.000 000 001	10^{-9}	–90	40	30

The decibel scale can be used to express the ratio of any two quantities, provided such quantities are expressed in terms which have a linear relationship to power. (Table A6.2).

TABLE A6.2 Sound power : examples

Source	*Sound power*	*Sound power level (dB re 10^{-12} W)*
Jet airplane	10 kW	160
Pneumatic chipping hammer	1 W	120
Automobile at 45 mph	0.1 W	110
Piano	20 mW	103
Conversational speech, connected	20 μW	73
Small electric clock	0.02 μW	43
Soft whisper	0.001 μW	30

Since sound power varies as square of sound pressure,

$$\text{pressure level in dB} = 10 \log_{10} \left(\frac{\text{sound pressure level}}{\text{reference pressure}}\right)^2 \text{, or}$$

$$\text{SPL (dB)} = 20 \log_{10} \left(\frac{p}{\text{reference pressure}}\right).$$

The accepted reference used for pressure is 0.0002 microbars or dynes per square centimetre; this is the threshold of hearing for average subjects.

It must be borne in mind that a decibel figure without any reference value is meaningless. In this guide, the notation 'dB' is used for both sound power and pressure level. Where not stated explicitly, the reference implied is 10^{-12} W for sound power level and 0.0002 dyne/cm^2 for sound pressure level.

A6.3 Combination of sound sources.

If it is required to add two absolute quantities, say, power in watts or pressure in microbars, then the total is obtained by simple addition. This is not possible, however, when the quantities are expressed in decibels; the quantities expressed in decibels are combined rather than added.

Table A6.3 shows the corrections to be applied to the higher level of the two for various differences between the two levels to be combined. For example, if it is required to combine two levels each 20 dB, then for 0 difference the correction factor is 3, and so the total level resulting will be 23 dB.

For all practical purposes, if the difference between two levels exceeds 9 dB, then the effect of the lower on the higher can be taken as negligible.

TABLE A6.3 Combination of SPLs

Difference between SPLs	*Add to larger SPL*
0, 1	+3
2, 3	+2
4, 5, 6, 7, 8, 9	+1
10 or more	0

A6.4 Presentation of velocity generated noise data : Strouhal number.

The data for velocity generated noise of duct components is usually presented in a generalised form. This is conveniently done by expressing the noise data as a function of the Strouhal number.

Strouhal number is a dimensionless ratio of the product of frequency and duct diameter to the velocity of air, i.e.

$$N_{STR} = \frac{FD}{V}$$

where
F is frequency of sound
D is duct diameter
V is velocity of air
} in compatible units

FIG A6.1 Relative responses for A, B and C weighting networks

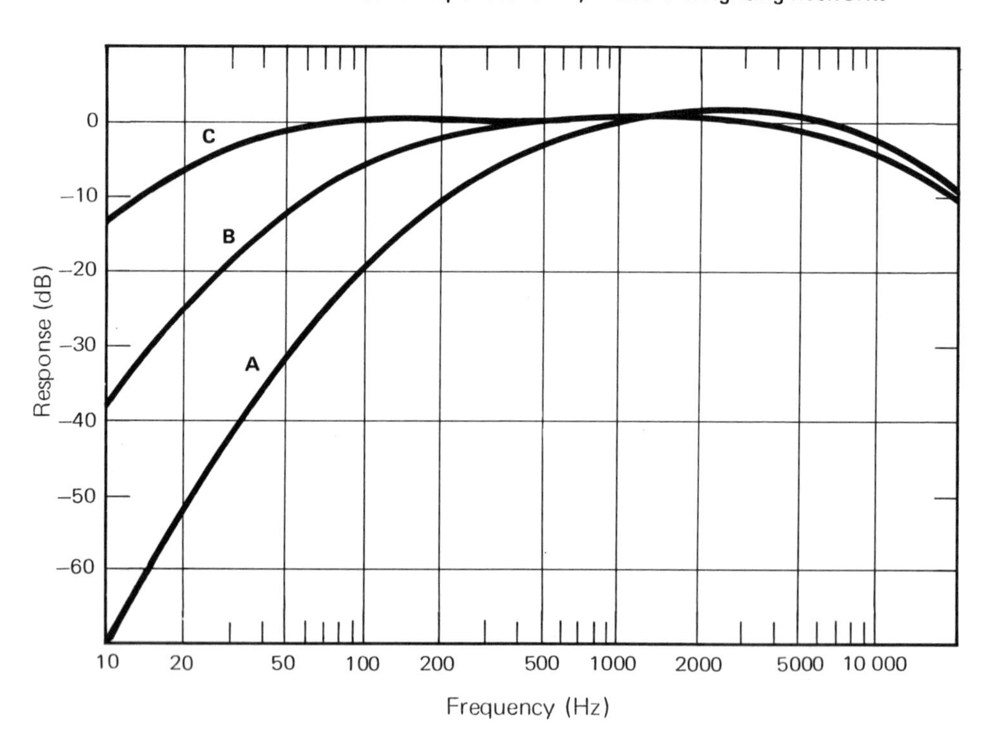

A6.5 Weighting networks.

The human ear has a set of highly sensitive networks of analysing filters. These enable the ear to have different responses to sounds of varying frequencies and intensities. For sound pressure levels up to 55 dB, the ear is less sensitive to low frequencies and more sensitive to the higher end of the spectrum.

Between sound pressure levels of 55 and 85 dB, the response curve flattens out and, at about 85 dB and above, the response curve of the ear is reasonably flat. To reproduce earlike conditions in sound pressure level meters, electronic filter networks are incorporated; commonly known as weighting networks A, B and C, their magnitudes are shown in Fig. A6.1. With weighting networks, it is possible to measure sound pressure levels which would be related closely to the subjective perception of loudness.

Sound pressure levels therefore measured with, for instance an A weighting network introduced into the sound level meter are given as so many dBA. The B and C weighting networks have these days fallen out of use and the A weighting network has achieved general acceptance over the whole range. An increase or decrease of 10 dBA corresponds roughly to doubling or halving of loudness. Thus 80 dBA is twice as loud as 70 dBA.

A6.6 Spectrum.

Weighting networks provide a simple way of expressing a certain sound as a single figure. Although this approximates to the perception of loudness, it does not give a complete picture of the exact nature of that sound. For analytical work, it is necessary to have a detailed knowledge of sound levels in each frequency band; this is known as the sound level spectrum. To obtain such a spectrum, analysing filters are introduced between the microphone and the indicating meter. The purpose of these filters is to isolate a particular band of frequencies, allowing them to pass freely to the measuring instrument, while blocking all other frequencies.

A6.7 The octave band.

When the audible frequency range is divided into eight bands, the upper frequency of each band being taken as twice that of the lower end, each band is called an octave band (see Table A6.4).

TABLE A6.4

Mid-octave Hz	63	125	250	500	1000	2000	4000	8000 Hz
Band or octave no.	1	2	3	4	5	6	7	8

A6.8 Noise criteria curves.

In 1955 Beranek published his Noise Criteria curves which are now well known as NC curves. These put single value contours on sound pressure level spectra and indicated the upper limits of acceptance for various environments. Figure A6.2 shows the NC curves.

Tabulated in Table A6.5 are the sound pressure levels at the various frequencies corresponding to the NC curves illustrated in Fig. A6.2. The values in the tabular form are more convenient when performing calculations.

TABLE A6.5 Sound pressure levels (dB re 0.0002 dyne/cm^2)

NC	63	125	250	500	1000	2000	4000	8000 Hz
65	80	75	71	68	66	64	63	62
60	77	71	67	63	61	59	58	57
55	74	67	62	58	56	54	53	52
50	71	64	58	54	51	49	48	47
45	67	60	54	49	46	44	43	42
40	64	57	50	45	41	39	38	37
35	60	52	45	40	36	34	33	32
30	57	48	41	35	31	29	28	27
25	54	44	37	31	27	24	22	21

FIG A6.2 Noise Criteria curves

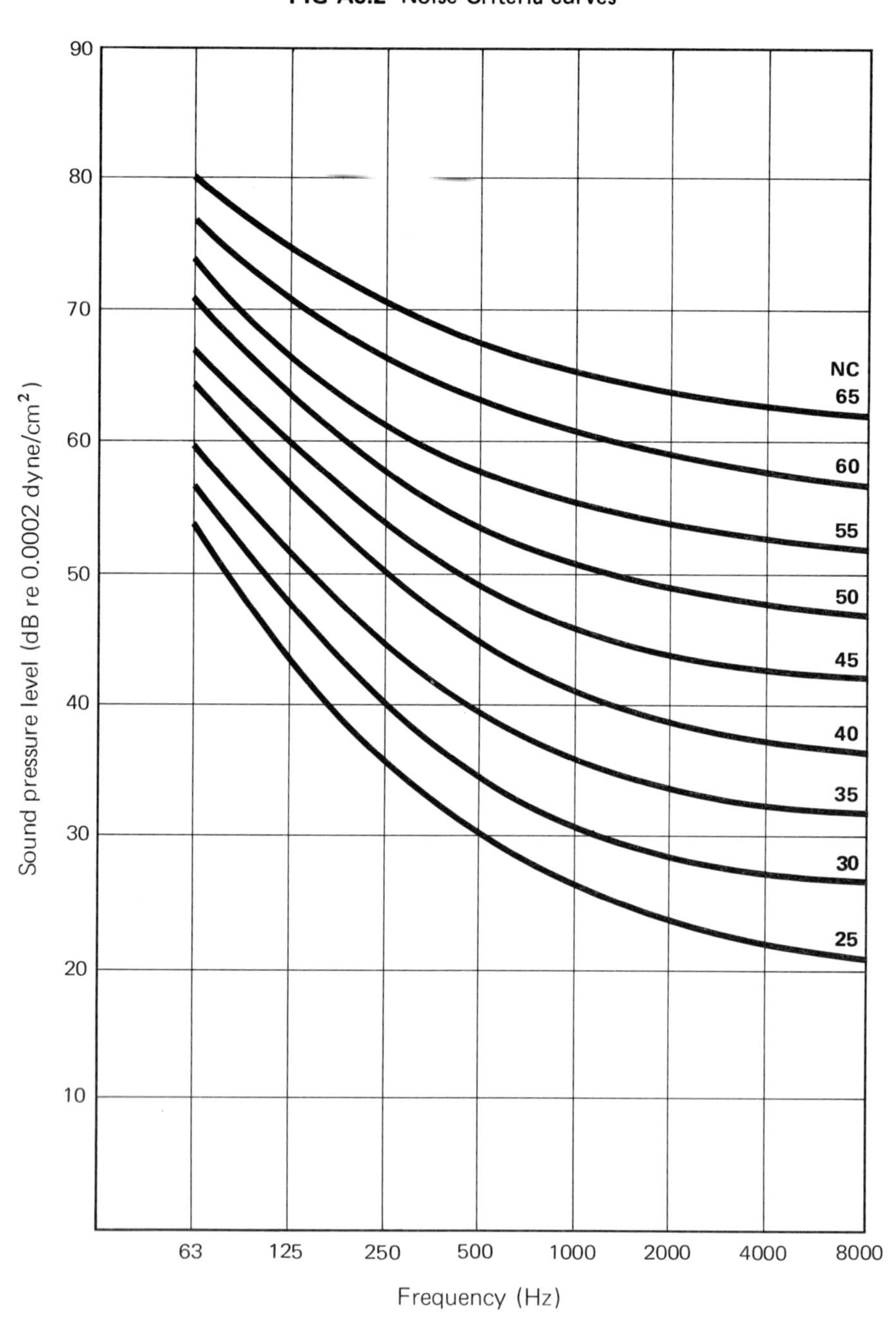

FIG A6.3 Noise Rating curves

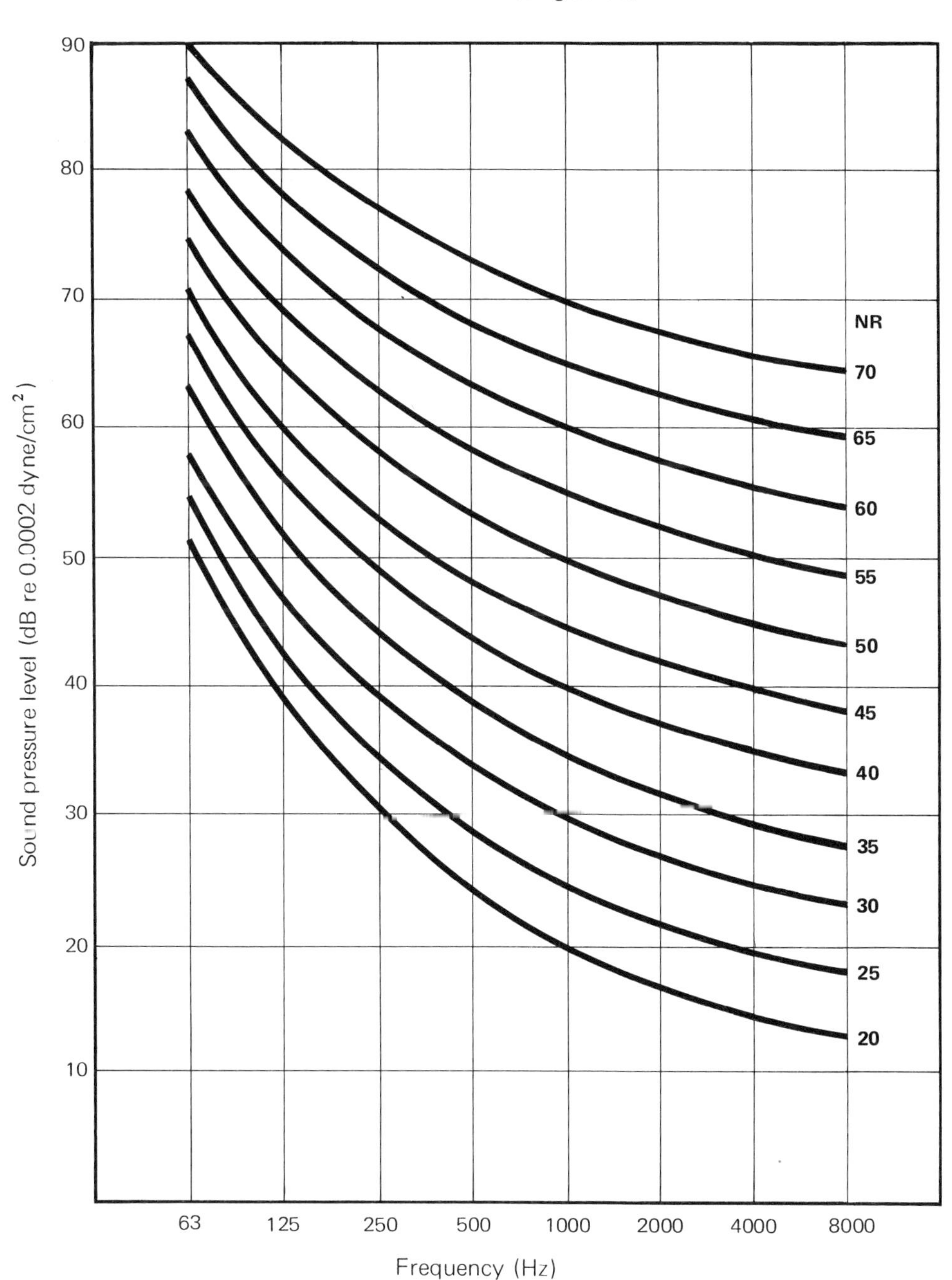

A6.9 Noise rating curves.

Noise rating (NR) curves are not in general use at present in the heating and ventilating industry, but are being adopted by the International Standards Organisation (ISO) as the recommended curves (Fig. A6.3). These are slightly different to the NC curves and relate to the sound pressure level in the 1000 Hz octave band.

A6.10 Sound power level and sound pressure level.

Although these two quantities have been defined earlier, it is well worth elaborating on the relationship between them. A considerable amount of confusion generally exists in their relative use.

There is an analogy in that the measurement of sound pressure level is comparable to the measurement of temperature in a room, whereas the sound power level is equivalent to the cooling capacity of the plant conditioning the room.

The resulting temperature is a function of the cooling capacity of the plant and the heat gains and losses of the room. Exactly in the same way the resulting sound pressure level would be a function of sound power output of the equipment together with the acoustic properties of the space.

Given the total sound power output of a plant and knowing the acoustic properties and dimensions of the space, it is possible to predict the sound pressure levels produced.

A6.11 Direct and reverberant sound pressure levels.

The sound pressure level in a room due to a noise source in that room is made up of two parts, sound travelling directly from the source to the listener's ear, and sound reaching the listener's ear after reflections from wall, floors and ceilings (Fig. A6.4).

The former is called the 'direct' and the latter the 'reverberant' sound pressure level. The direct level depends on the distance between the source and the listener and falls off rapidly with distance (as the square of distance to be exact). This level therefore dominates close to the source.

The reverberant level, on the other hand does not depend on distance, but on the size and absorption characteristics of the room. The total sound pressure level is the logarithmic summation of the two. (See Figs. A6.4 and A6.5.)

FIG A6.4 Direct and reflected (reverberant) sound

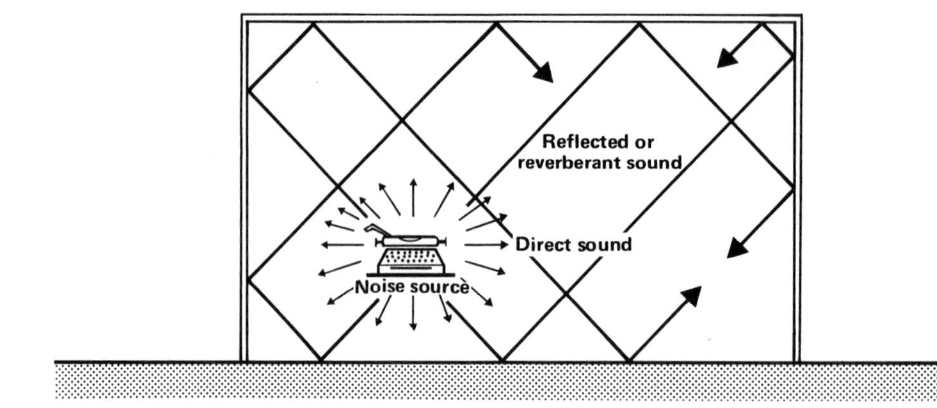

FIG A6.5 Effect of distance from source on direct and reverberant sound pressure levels in a space

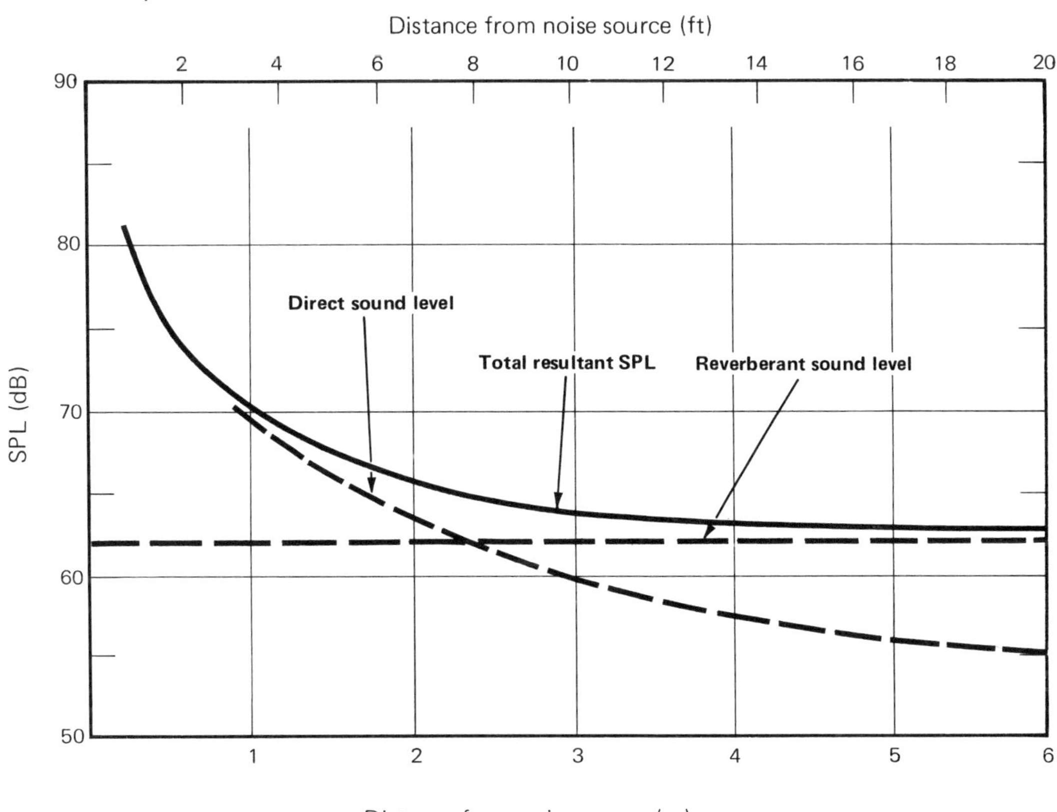

A6.12 Room characteristics (room absorption).

This parameter basically specifies the ability of the room to absorb noise. It can be expressed either in terms of its reverberation time and volume or alternatively absorption coefficient and total surface area of walls, floor and ceilings.

Reverberation time

This is the time required for the average sound pressure level in the room to decrease by 60 dB after the noise source has stopped generating.

Absorption coefficient

This coefficient is the ratio of the sound energy absorbed by a surface to the energy incident upon it. Absorption coefficients of some common building materials are shown in Appendix 2. Rooms having a longer reverberation time or a smaller absorption coefficient are usually termed as 'live'. Those with opposite characteristics are called 'dead'. A reverberation time of 1 second is about average.

A6.13 Acoustic materials.

The most important piece of knowledge required about acoustic materials is a realisation that there are two kinds and that they are in no way substitutes for each other. They are insulating materials and absorbing materials respectively.

Insulating materials

Insulating materials restrict the transmission of sound through themselves. A typical example is a brick wall, a less effective example is the metal wall of a duct. Note that these materials tend to be airtight and rather heavy. In fact a useful 'rule of thumb' for checking insulation is first to check whether it is airtight (remember small holes can have serious effects) and then to check the weight per square foot (superficial density) at the lightest point, the densest materials usually giving the greatest insulation, sheet lead being particularly effective.

It is worth noting that although insulating materials restrict the passage of sound they do not make it disappear. It is usually simply reflected back in roughly the same way as a mirror reflects light and this often causes a build up of sound on the source side.

Absorbent materials

Absorbent materials do not reflect sound (or only reflect a small proportion of it). Typical absorbers are mineral wool and open windows. When sound falls on these it does not return. In the case of mineral wool and similar materials some of it tends to be lost in the interstices of the material, but in the case of open windows it simply passes through. In both cases a high proportion of the sound passes through to the other side. It is therefore obvious that absorbent materials are not the same as insulating materials and cannot be used as such.

A6.14 Silencers.

Silencers are of special interest to heating and ventilating engineers because they allow the passage of air while restricting the passage of sound. They usually subdivide the airflow into several passages each lined with perforated sheet backed by mineral wool, glassfibre or some other sound absorbing material.

Silencers are generally specified by the sound attenuation in decibels they provide in each octave band, so that the degree of match with the sound power distribution of the noise source over the frequencies may be judged. This is known as the insertion loss of the silencer.

The other important parameter associated with silencers is the resistance to airflow. It would clearly be unsatisfactory to introduce so much resistance against the fan (in order to absorb the noise) that the fan speed had to be increased, thereby generating more sound and incurring additional power consumption.

With the ever increasing air velocities in modern airconditioning systems, it has also become necessary to take into account the noise generated by the turbulence created due to the presence of the silencer in the air stream. This is known as the self noise of the silencer, and is discussed in greater detail in Chapter 3 on velocity generated noise.

Correct location of the silencers in the system is of great importance in order to ensure their maximum effectiveness, see Figs. 5.3 and 5.4 (page 66).

A6.15 Vibration and its isolation.

All rotating and reciprocating machinery transmits vibrations of a definite frequency (usually 50 Hz or lower) and low frequency noise to the structure to which they are rigidly fixed.

Anti-vibration mountings are, therefore, normally necessary between the machine and the structure if vibration and noise are to be avoided. These mountings can be springs, rubber or any resilient material, excluding cork, provided its natural frequency is well below the lowest rotating or oscillating frequency of the machine. In the event of these frequencies being close, resonance could occur which can result in dangerously large amplitudes (amplification of vibrations) and may damage the structure and the machine, see Fig. A6.6. Careful selection of the mountings is therefore essential. *The natural frequency of resilient material is not solely dependent upon its static deflection under load,* as is commonly believed, and it is important to obtain detailed information from the manufacturer of the mount.

Figure A6.6 shows the relative amplitude of the structure which, in the absence of any damping, can become infinitely large at resonance, i.e. when natural frequency of the mount equals that of the source of vibrations. It can be seen that in order to reduce the relative amplitude of the structure adequately (i.e. achieve substantial vibration isolation), the natural frequency of the mount should be less than one-third of the natural frequency of the source.

A6.16 Transmission loss (TL) or sound reduction index (R).

The measure of airborne sound insulation of a partition is known as either transmission loss (TL) or sound reduction index (R), the former being preferred in the USA while the latter is essentially British. In either case it is equal to the number of decibels by which sound incident upon a partition is reduced during transmission through it.

It is important to note that figures of TL quoted by the manufacturers of the partition materials are laboratory test figures. The installed transmission loss figures, however, are often 5-10 dB lower. This is due to the fact that in actual installations perfect air seals are very rarely achieved.

FIG A6.6 Effect of frequency ratio on amplitude for undamped oscillations

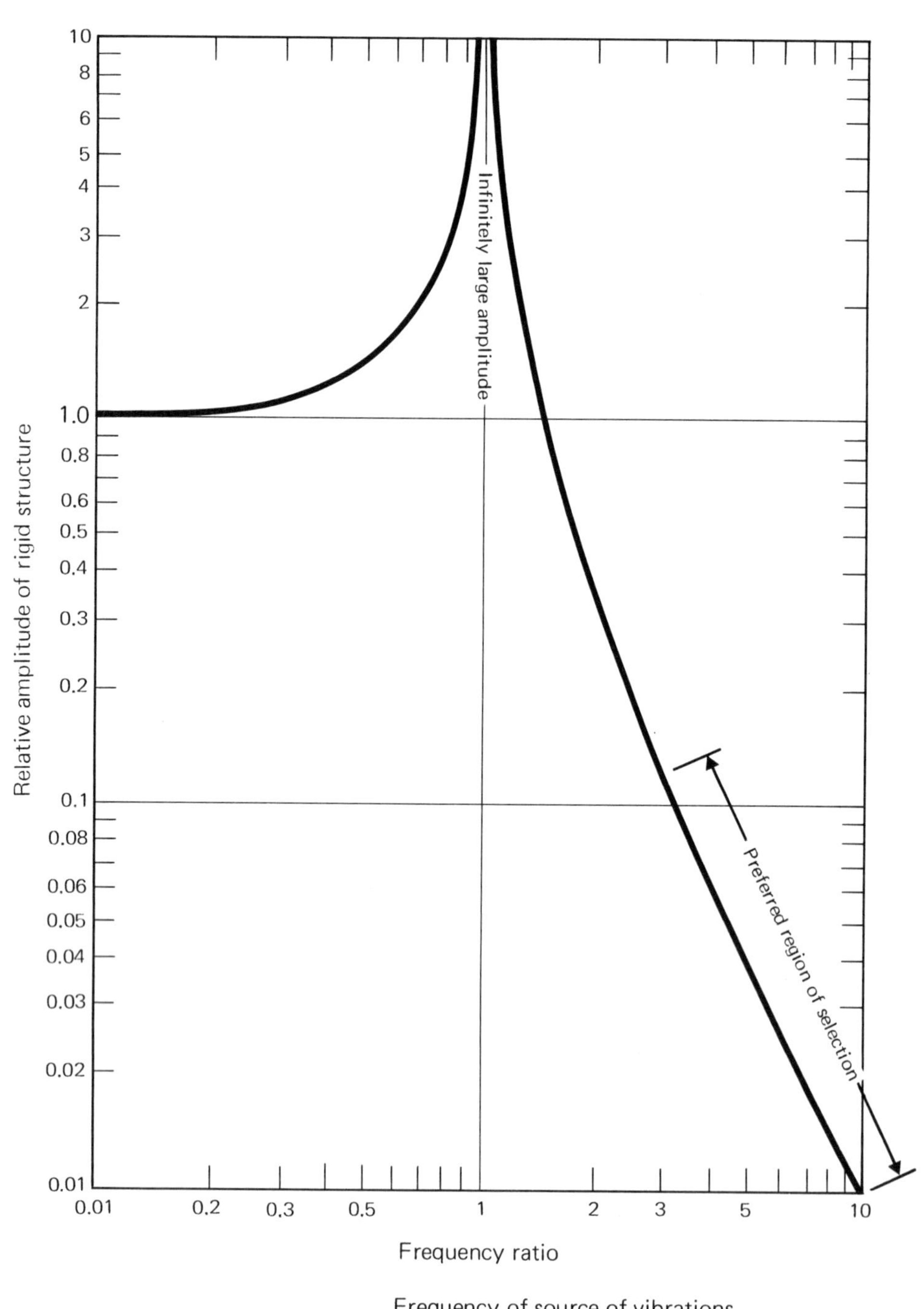

$$\text{Frequency ratio} = \frac{\text{Frequency of source of vibrations}}{\text{Natural frequency of A/V mount}}$$

APPENDIX 7

Suggestions for further reading

ASHRAE, *Ashrae Handbook, 1973, Systems Volume, chapter 35.*
Beranek, L.L. (ed.), *Noise Reduction,* McGraw-Hill, New York, 1960.
Harris, C.M. (ed.), *Handbook of Noise Control,* McGraw-Hill, New York, 1957.
Jorgensen, R. (ed.), *Fan Engineering,* Buffalo Forge Co., New York, 1961.
Parkin, P.H. and Humphreys, H.R., *Acoustics, Noise and Buildings,* Faber, London, 1969.
Sharland, I., *Woods Practical Guide to Noise Control,* Woods of Colchester, Colchester 1972.
Woods of Colchester, *Design for Sound,* Woods of Colchester, Colchester.
Woods, R. I., *Noise Control in Mechanical Services,* Sound Research Laboratories, Colchester 1972.

INDEX